Land Use Planning and Management for a Sustainable Future

Edited by Masa Noguchi

Published in London, United Kingdom

Land Use Planning and Management for a Sustainable Future
http://dx.doi.org/10.5772/intechopen.1008127
Edited by Masa Noguchi

Contributors
Abdelrahim Salih, Ahmed Abdalla, Anca Crăciun, Angheluță Vădineanu, Anwarelsadat Elmahal, Etaib Saeed Ganawa, Jhon Jairo Castañeda Fandiño, Jianqiang Huang, Ozge Celik Yilmaz, Patrick O'Connor, Radu Ştefan Vădineanu, Rattana Boonprasert, Wifag Hassan Mahmoud

First published in London, United Kingdom, 2026 by IntechOpen
IntechOpen is the global imprint of INTECHOPEN LIMITED, registered in England and Wales, registration number: 11086078, 167-169 Great Portland Street, London, W1W 5PF, United Kingdom

For EU product safety concerns: IN TECH d.o.o., Prolaz Marije Krucifikse Kozulić 3, 51000 Rijeka, Croatia, info@intechopen.com or visit our website at intechopen.com.

British Library Cataloguing-in-Publication Data
A catalogue record for this book is available from the British Library

Land Use Planning and Management for a Sustainable Future
Edited by Masa Noguchi
p. cm.

This title is part of the Sustainable Development Book Series, Volume 45
Topic: Urban Planning and Environmental Management
Series Editor: Usha Iyer-Raniga
Topic Editors: Rui Alexandre Castanho and Luís Loures

Print ISBN 978-1-83635-656-1
Online ISBN 978-1-83635-655-4
eBook (PDF) ISBN 978-1-83635-657-8
ISSN 2753-6580

If disposing of this product, please recycle the paper responsibly.

IntechOpen Book Series
Sustainable Development
Volume 45

Aims and Scope of the Series

Transforming our World: the 2030 Agenda for Sustainable Development endorsed by United Nations and 193 Member States, came into effect on Jan 1, 2016, to guide decision making and actions to the year 2030 and beyond. Central to this Agenda are 17 Goals, 169 associated targets and over 230 indicators that are reviewed annually. The vision envisaged in the implementation of the SDGs is centered on the five Ps: People, Planet, Prosperity, Peace and Partnership. This call for renewed focused efforts ensure we have a safe and healthy planet for current and future generations.

This Series focuses on covering research and applied research involving the five Ps through the following topics:

1. Sustainable Economy and Fair Society that relates to SDG 1 on No Poverty, SDG 2 on Zero Hunger, SDG 8 on Decent Work and Economic Growth, SDG 10 on Reduced Inequalities, SDG 12 on Responsible Consumption and Production, and SDG 17 Partnership for the Goals

2. Health and Wellbeing focusing on SDG 3 on Good Health and Wellbeing and SDG 6 on Clean Water and Sanitation

3. Inclusivity and Social Equality involving SDG 4 on Quality Education, SDG 5 on Gender Equality, and SDG 16 on Peace, Justice and Strong Institutions

4. Climate Change and Environmental Sustainability comprising SDG 13 on Climate Action, SDG 14 on Life Below Water, and SDG 15 on Life on Land

5. Urban Planning and Environmental Management embracing SDG 7 on Affordable Clean Energy, SDG 9 on Industry, Innovation and Infrastructure, and SDG 11 on Sustainable Cities and Communities.

The series also seeks to support the use of cross cutting SDGs, as many of the goals listed above, targets and indicators are all interconnected to impact our lives and the decisions we make on a daily basis, making them impossible to tie to a single topic.

Meet the Series Editor

Usha Iyer-Raniga is a professor in the School of Property and Construction Management at RMIT University. Usha co-leads the One Planet Network's Sustainable Buildings and Construction Programme (SBC), a United Nations 10 Year Framework of Programmes on Sustainable Consumption and Production (UN 10FYP SCP) aligned with Sustainable Development Goal 12. The work also directly impacts SDG 11 on Sustainable Cities and Communities. She completed her undergraduate degree as an architect before obtaining her Masters degree from Canada and her Doctorate in Australia. Usha has been a keynote speaker as well as an invited speaker at national and international conferences, seminars and workshops. Her teaching experience includes teaching in Asian countries. She has advised Austrade, APEC, national, state and local governments. She serves as a reviewer and a member of the scientific committee for national and international refereed journals and refereed conferences. She is on the editorial board for refereed journals and has worked on Special Issues. Usha has served and continues to serve on the Boards of several not-for-profit organisations and she has also served as panel judge for a number of awards including the Premiers Sustainability Award in Victoria and the International Green Gown Awards. Usha has published over 100 publications, including research and consulting reports. Her publications cover a wide range of scientific and technical research publications that include edited books, book chapters, refereed journals, refereed conference papers and reports for local, state and federal government clients. She has also produced podcasts for various organisations and participated in media interviews. She has received state, national and international funding worth over USD $25 million. Usha has been awarded the Quarterly Franklin Membership by London Journals Press (UK). Her biography has been included in the Marquis Who's Who in the World® 2018, 2016 (33rd Edition), along with approximately 55,000 of the most accomplished men and women from around the world, including luminaries as U.N. Secretary-General Ban Ki-moon. In 2017, Usha was awarded the Marquis Who's Who Lifetime Achiever Award.

Meet the Volume Editor

Dr. Masa Noguchi is an Associate Professor in Environmental Design at the Faculty of Architecture, Building and Planning, University of Melbourne. He focuses on Environmental Experience Design (EXD) and mass customization strategies to improve energy efficiency, affordability, occupant well-being, and community development in the built environment. As the founding coordinator of the ZEMCH Network, he leads global research on zero-energy mass-customized homes (ZEMCH) and drives vertical subdivision development projects internationally. A Chartered Engineer, environmentalist, and technological product designer, Dr. Noguchi designed an award-winning near-net-zero-energy housing project in Canada and contributed to sustainable community development in Scotland. He actively promotes sustainable solutions through teaching, workshops, and international conferences, bridging research, design, and practical implementation to create resilient, energy-efficient, and people-centered built environments.

Contents

Preface

Land use planning and management lie at the heart of sustainable development, shaping how communities grow, natural resources are used, and ecosystems are preserved. In an era marked by rapid urbanization, climate change, and increasing environmental pressures, the need for thoughtful, strategic approaches to land use has never been more urgent. This book, *Land Use Planning and Management for a Sustainable Future*, seeks to provide a comprehensive exploration of how integrated planning, policy interventions, and participatory governance can guide land use toward sustainability.

The central premise of this work is that sustainable land use cannot be achieved through fragmented or sectoral approaches alone. Urban development, agriculture, forestry, transportation, and environmental conservation are deeply interconnected, and decisions in one domain often have cascading effects on others. Recognizing this complexity, the book emphasizes the importance of systems thinking in land use planning, highlighting the need to balance economic growth, social equity, and ecological integrity.

Drawing on extensive case studies, policy analyses, and best practices from diverse geographic contexts, this book examines the tools, strategies, and institutional frameworks that enable effective land use management. It explores traditional planning approaches alongside innovative concepts such as smart growth, green infrastructure, ecosystem-based management, and community-led spatial planning. In doing so, it underscores the critical role of evidence-based decision-making, stakeholder participation, and adaptive governance in creating resilient and sustainable landscapes.

A key theme of the book is the interplay between policy and practice. While regulations, zoning laws, and master plans provide essential guidance, implementing these policies often depends on local capacity, community engagement, and inter-agency coordination. Through examples ranging from urban redevelopment projects to watershed management initiatives, the book illustrates how participatory approaches and collaborative governance can transform planning frameworks into effective, actionable strategies for sustainability.

This work also addresses the emerging challenges posed by climate change, technological advancements, and demographic shifts. It highlights the roles of spatial data, geographic information systems (GIS), and decision-support tools in anticipating future land-use scenarios and informing adaptive strategies. By integrating these technological and analytical tools with human-centered planning approaches, the book provides a holistic framework for managing land resources that is both forward-looking and contextually grounded.

It is the hope of this book that *Land Use Planning and Management for a Sustainable Future* will serve as a resource for scholars, practitioners, policymakers, and students interested in advancing sustainable land use practices. By blending theory, empirical evidence, and practical guidance, the book aims to inspire informed decision-making and responsible stewardship of land resources, ultimately contributing to a future where development and environmental sustainability coexist in harmony.

Masa Noguchi
Faculty of Architecture, Building and Planning,
The University of Melbourne,
Melbourne, Australia

Chapter 1

Sustainable Agriculture, Climate Change, and Remote Sensing: A Review

Jhon Jairo Castañeda Fandiño

Abstract

Sustainable agriculture has become a critical framework for addressing the escalating challenges posed by climate change, food security, and environmental degradation. This review article provides a comprehensive synthesis of recent advances linking sustainable agricultural practices and their adaptation to and mitigation of climate change impacts, with a particular focus on the role of remote sensing technologies. We review the integration of remote sensing and artificial intelligence (AI) tools in monitoring land use and land cover changes, crop growth dynamics, soil organic carbon status, and greenhouse gas emissions in agricultural landscapes. The findings reveal that remote sensing, combined with advanced machine learning models, significantly enhances the spatial and temporal precision of agricultural monitoring, enabling improved decision-making for sustainable management. However, persistent challenges remain, including data gaps for smallholder farms, regional disparities in technology adoption, and the need for multi-factorial, integrative models that incorporate socioeconomic variables alongside climatic drivers. This article underscores the imperative of deploying scalable, cost-effective remote sensing data coupled with AI to foster climate-smart agriculture, emphasizing policy implications and future research directions to bridge existing knowledge gaps and improve resilience against climate variability.

Keywords: sustainable agriculture, climate change, remote sensing, machine learning, land use and land cover, soil organic carbon, greenhouse gas emissions, precision agriculture

1. Introduction

Sustainable agriculture represents an integrated approach to food production that balances the goals of environmental stewardship, economic viability, and social equity. Amid the profound and accelerating impacts of climate change, including heightened temperature variability, altered precipitation patterns, and increased frequency of extreme weather events, agriculture faces unprecedented challenges that threaten global food security and ecosystem resilience. Agriculture itself contributes significantly to climate change through greenhouse gas (GHG) emissions, water usage, and land cover changes, necessitating adaptive strategies that can simultaneously sustain productivity and mitigate environmental impacts [1].

Agriculture stands at the crossroads of addressing global food security and mitigating environmental impacts amid an increasingly volatile climate system. Climate change has introduced pronounced variability in temperature and precipitation patterns, leading to stressful conditions for crop growth and threatening the sustainability of food systems worldwide. Agricultural activities have a dual role: they are both vulnerable to climate-induced alterations and significant contributors to environmental degradation, notably through land-use changes and GHG emissions. Studies have underscored that global agricultural production systems are intertwined with water and energy consumption and contribute substantially to GHGs, complicating the nexus between food security and environmental sustainability [2].

Sustainable agriculture has evolved as a conceptual and operational framework aimed at balancing productivity with environmental conservation. Strategies focus on enhancing resource-use efficiency, conserving biodiversity, and promoting farming practices resilient to climatic shocks. The urgency of sustainable agriculture is accentuated in developing regions, where smallholder farms predominate and are disproportionately affected by climate variability. Innovations in agricultural technology, including climate-resilient crop varieties and integrated pest management, form part of these adaptive strategies but often fail to reach marginalized farming communities effectively [3].

Technological advancements in remote sensing and AI present promising avenues to address some of these challenges. Remote sensing offers spatially and temporally continuous data, critical for understanding land use dynamics, monitoring crop health, and quantifying water and carbon fluxes at scales ranging from field plots to entire regions. Coupled with AI and ML, these technologies enable sophisticated analyses such as crop classification, drought forecasting, and GHG emissions estimation, facilitating data-driven decision-making for climate-smart agriculture. The integration of such tools in agricultural monitoring systems is pivotal in adapting practices to the multifaceted threats posed by climate change [4].

Agricultural systems face multifactorial stressors intensified by climate change. Land degradation through soil erosion, nutrient depletion, and unsustainable farming practices intensifies productivity losses and reduces the land's resilience capacity. Water scarcity, driven by altered precipitation regimes and increased evapotranspiration due to rising temperatures, further constrains agricultural outputs. Moreover, agricultural activities contribute significantly to GHG emissions, especially methane from flooded rice paddies and nitrous oxide from fertilized soils, exacerbating the climate crisis [5].

Addressing these issues demands holistic, data-driven approaches that integrate multiple environmental and socioeconomic variables. Conventional monitoring approaches, relying on ground-based observations, often lack spatial coverage and are insufficient for resolving complex agricultural-climate interactions. There is an imperative to harness remote sensing data combined with ML algorithms to improve monitoring capabilities and develop predictive models that inform sustainable resource management, promoting the resilience and sustainability of agroecosystems [6].

Remote sensing technologies, particularly when combined with advanced data analytics such as artificial intelligence (AI) and machine learning (ML), have emerged as powerful tools for monitoring and managing agricultural systems under changing climatic conditions. Satellite-based sensors provide spatially explicit, temporal, and cost-efficient data across varied scales, enabling detailed assessments of crop health, soil properties, land-use changes, and environmental stressors. Despite rapid technological advancements, challenges persist in integrating these tools effectively into sustainable agriculture frameworks, particularly for smallholder farmers in developing countries, where data scarcity and knowledge gaps limit the translation of remote sensing insights into actionable practices [7].

This review consolidates recent scientific findings on sustainable agriculture in the context of climate change, emphasizing the critical role of remote sensing and machine learning applications. We examine methodologies, discuss thematic trends, analyze challenges, and identify research gaps to inform future policy and scientific endeavors.

2. Review methodology

This review was conducted through a systematic search and analysis of academic literature from multiple databases, including Scopus, Web of Science, IEEE Xplore, and MDPI Journals, up to mid-2025. Search terms incorporated combinations of "sustainable agriculture," "climate change," "remote sensing," "machine learning," "land use change," "greenhouse gas emissions," and related keywords. Studies were included if they focused on empirical or modeling research integrating remote sensing technologies and AI/ML techniques with sustainable agricultural goals, particularly those addressing climate change adaptation or mitigation.

Inclusion criteria required publication within the last five years to reflect recent advances, a focus on agricultural systems, and availability of full text in English. Exclusion criteria involved studies without data-driven analysis or those unrelated to agricultural or environmental sustainability. The initial search yielded over 250 publications, which were screened for relevance based on titles and abstracts. Following this, 75 studies qualified for full-text review. A final curated set of 50 key papers was selected based on methodological rigor, geographic and thematic diversity, and citation impact.

The literature selection process is summarized in **Figure 1**.

3. Results and analysis

3.1 Sustainable agriculture and climate change interactions

Climate change exacerbates vulnerabilities in agricultural systems by altering water availability, increasing the frequency of droughts, and introducing new pest and disease dynamics. Agriculture contributes significantly to GHG emissions, particularly methane from rice paddies and nitrous oxide from synthetic fertilizers, underscoring its centrality in climate mitigation efforts. The concept of the water–energy–food–carbon (WEFC) nexus is gaining prominence,

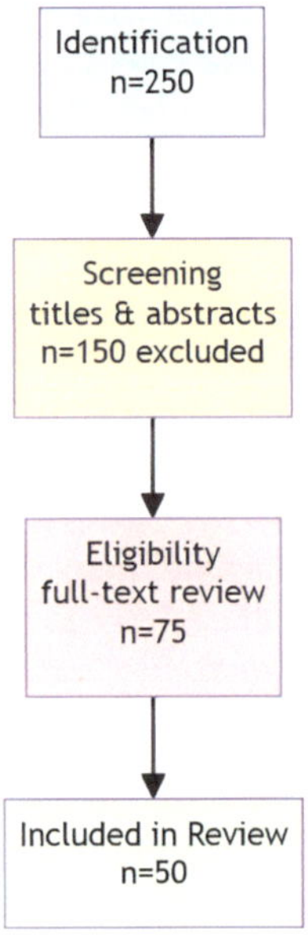

Figure 1.
Selection diagram in literature selection.

highlighting the interdependence of resource use in sustainable agriculture and the urgent need for integrated management strategies [8].

Research reveals that sustainable agriculture must incorporate climate-resilient management practices, including diversification of crops, conservation tillage, and improved nutrient use efficiency. However, empirical evidence suggests that technology transfer to smallholder farmers remains limited, often referred to as the "Death Valley of impact" due to fragmented research approaches, lack of community engagement, and supply-driven technology dissemination rather than demand-responsive frameworks [3].

3.2 Role of remote sensing in agricultural monitoring

Remote sensing has become indispensable for monitoring land use and land cover (LULC) changes that affect agricultural sustainability and environmental health. Studies in South Punjab, Pakistan, demonstrate how population growth drives LULC dynamics, impacting agricultural land and contributing to climate change pressures. High-resolution satellite data, combined with geographic information system (GIS) tools, provide precise detection of these changes over time [3].

Advancements in synthetic aperture radar (SAR) and multispectral sensors enable improved crop type classification, growth stage monitoring, and biomass estimation, even under challenging weather conditions such as frequent cloud cover typical of tropical regions [4]. SAR data, such as Sentinel-1, combined with optical sensors such as PROBA-V, have facilitated improved rice crop monitoring by integrating backscatter profiles and vegetation indices, offering near-real-time insights critical for managing climate risks [4].

Further, global assessments utilizing Google Earth Engine (GEE) platforms have accelerated the processing and analysis of multi-temporal remote sensing data over large spatial extents, enhancing the capacity to monitor crop expansion and contraction and their associated environmental impacts [5].

3.3 ML and AI in precision agriculture

The integration of ML with remote sensing data has revolutionized the precision agriculture field, enabling improved prediction accuracy for crop yields, water requirements, and soil conditions. Convolutional neural networks (CNNs) and attention-based deep learning models show superior performance in land cover classification tasks by effectively capturing the spatial and temporal heterogeneity of agricultural landscapes [6].

Emerging platforms, such as Farmdee-Mesook, employ AI-driven crop modeling combined with satellite data to generate site-specific agronomic recommendations, showing measurable benefits in yield increases and input cost reductions among smallholder rice farmers in Southeast Asia [7]. These innovations demonstrate the potential for AI tools to translate complex data into actionable strategies, fostering sustainability and resilience.

In addition, machine learning models are critical in monitoring soil organic carbon (SOC), a key indicator of soil health and carbon sequestration potential. By combining soil sampling data with multisensor remote sensing inputs, ML algorithms can spatially estimate SOC variations with improved accuracy, facilitating climate-smart land management practices [8]. Similar approaches also extend to estimating soil moisture and evapotranspiration for water management optimization [3].

3.4 Greenhouse gas emissions and climate mitigation

Linking agricultural expansion and intensification with GHG emissions has been a primary research focus. Remote sensing-based studies quantify how agricultural area expansion, particularly in tropical regions, accelerates carbon dioxide and nitrous oxide emissions, thus guiding mitigation policies [1]. Furthermore, irrigation practices and water management strategies critically influence methane emissions from flooded rice paddies, often the primary source of agricultural methane [9].

Smart agriculture technologies offer promising avenues to reduce GHG emissions by optimizing input use, particularly water and fertilizers, through site-specific applications enabled by IoT sensors and remote data analytics [4]. However, scaling these technologies remains challenging, especially in regions with limited infrastructure and small-scale farming operations.

3.5 Challenges and limitations

Despite significant progress, several challenges impede the full realization of remote sensing and AI for sustainable agriculture under climate change stressors. Data limitations, including cloud cover interference and temporal resolution constraints, persist, particularly in tropical and developing regions. Many crop monitoring systems are designed for specific geographies and lack portability, limiting their utility in diverse agricultural contexts [10].

Table 1 summarizes a broad set of interdisciplinary insights focused on the intersection between water sustainability, climate change, agricultural innovation,

Main topic	Subtopic/category	Key contributions	References/citations
Global hydrological cycle and water sustainability	Groundwater depletion and management	Use of GRACE satellites to monitor aquifer depletion in India, China, and other Asian regions. Identification of significant rates of decline in groundwater storage, primarily due to agricultural overexploitation. These measurements complement in-situ data and hydrological models, improving regional accuracy.	[11, 12]
	Non-stationarity of the terrestrial hydrological cycle	Evidence of long-term changes and increasing variability in the hydrological cycle due to human activities and climate change is altering historical water availability patterns and affecting sustainable water management, with implications for infrastructure and early warning systems.	[11]
	Integration of remote sensing and AI in water resources assessment	Development of methodologies combining satellite data with artificial intelligence models to monitor drought trends, soil moisture levels, and agricultural water consumption. Application in regions such as California to optimize agricultural water use.	[1, 4, 11, 12]
Applications of AI and IoT in agriculture	Precision agriculture and smart greenhouse technologies	Demonstration of increased efficiency and productivity using IoT technologies combined with AI in greenhouses, along with barriers such as high initial costs, poor rural connectivity, and training gaps. Importance of training programs and subsidies to promote adoption among smallholder farmers.	[13, 14]
	Advanced image analysis and remote sensing for agricultural management	Proposal of deep learning architectures with self-attention mechanisms to improve crop classification and estimation of water consumption, even under cloudy conditions and spectral noise. Use of unlabeled data to enhance classification performance through self-supervised learning.	[1, 12–16]
	Ethical, security, and socioeconomic aspects	Identification of vulnerabilities in AI-based agricultural platforms, promoting the use of blockchain to enhance trust and data protection, addresses inequalities in technological access, the need for inclusive regulatory frameworks, and ethical protocols to ensure responsible deployment.	[4, 12–16]

Main topic	Subtopic/category	Key contributions	References/citations
Land use change and urbanization	Impact of population growth and urban expansion	Analysis of rapid urban area growth outpacing population growth, leading to the loss of agricultural land and challenges for food security. Observation of parallel trends of reforestation and improved land management in some regions, partially mitigating negative impacts.	[1, 14]
	Land use and land cover (LULC) classification and monitoring via multi-sensor integration	Improved classification accuracy through the fusion of multi-temporal, multi-spectral, and angular data. Development of methods combining deep neural network architectures with fusion operations to overcome limitations caused by cloud cover and spatial heterogeneity.	[17, 18]
Climate impacts on agricultural production and sustainability	Climate-induced patterns in crops and GHG emissions	Evidence of yield reductions due to rising temperatures and changing precipitation patterns. Assessment of the WEFC nexus, quantification of energy consumption in agriculture, and carbon and nitrous oxide emissions from agricultural expansion and fertilizer use. Linkage to sustainable policies for mitigation and adaptation.	[7, 8, 19]
	Adaptation strategies and policy recommendations	Importance of AI-based predictive models to anticipate extreme climate events and their integration into public policies to promote resilient, sustainable, and socially inclusive agricultural practices.	[20, 21]
	Irrigation practices and resource nexus	Evaluation of the cooling effect of irrigated crops and water-saving strategies, such as crop switching and improved irrigation efficiency, including market barriers. An integrated perspective of the water–energy–food nexus in irrigated agriculture.	[1, 4, 7, 8, 12–19, 22]
Machine learning models for agricultural prediction and monitoring	Hybrid approaches in yield prediction	Development of hybrid models combining algorithms such as RF, XGBoost, and LSTM, which improve accuracy in rice and other crop yield predictions by identifying key variables such as soil properties and climate.	[23]
	Soil organic carbon monitoring	Integration of machine learning techniques with remote sensing data to estimate SOC with high accuracy highlights the importance of feature selection and active learning in contrasting environments.	[1, 7, 8, 14, 17–24]
	New architectures and model optimization	Proposals for fused network architectures with attention mechanisms for remote sensing image classification, employing advanced optimization techniques to improve performance under data imbalance conditions. Application of semi-supervised learning to leverage unlabeled data.	[1, 4, 12, 14, 16–25]

Main topic	Subtopic/category	Key contributions	References/citations
Drought monitoring and forecasting with AI and remote sensing	Integrated indices and multi-sensor fusion methods	Development of composite drought indices integrating temperature, moisture, and precipitation data using techniques such as weighted averages and principal component analysis, improving spatial and temporal resolution across different regions.	[26]
	Deep learning models and forecasting	Creation of LSTM-based models with attention mechanisms for robust drought monitoring using multi-source data, applied in critical regions, with validation against meteorological and agricultural indices.	[21–28]
	Applications in unmonitored or data-scarce areas	Use of machine learning methods incorporating long-range climate forecasts to improve predictions in ungauged basins, facilitating regional drought management.	[1, 4, 7, 8, 12–28]
	Smart platforms for GHG monitoring and environmental management	Design of mobile platforms focused on smart agriculture, integrating simulated, satellite, and agronomic models to support smallholder farmers in managing and reducing GHG emissions. Use of accessible and user-friendly systems to increase adoption.	[29]
	Blockchain for data trust and security	Proposal of blockchain as a solution to enhance transparency, traceability, and data security in AI-based agricultural systems, addressing ethical and privacy concerns and facilitating sustainable practices.	[7, 8, 17–19]
	Multimodal analysis for water and nutrient management	Implementation of advanced preprocessing and optimization techniques for the accurate extraction of features related to crop water and nutrient requirements from remote sensing images, improving decision-making in irrigation and fertilization.	[4, 12–16]
Training, technology adoption, and public policies	Training and barriers to technology adoption	Identification of financial, infrastructure, and training barriers limiting the adoption of smart technologies in agriculture, especially for small producers. Emphasis on thematic training programs, subsidies, and advisory services to increase adoption rates and the replication of innovations.	[14]
	Pilot projects and international cooperation	Examples of multinational climate-smart agriculture projects supported by AI, addressing different crops and practices, and promoting comparative learning and technological standardization.	[13]
	Policy framework for sustainable and inclusive agriculture	Critical analysis of national policies promoting digital transformation in the agricultural sector highlights the need for inclusive frameworks with environmental assessments and circular economy principles for greater scalability and social impact.	[1, 3, 4, 7, 8, 12–29]

Main topic	Subtopic/category	Key contributions	References/citations
Future directions and research gaps	Technological gaps in model integration and generalization	Need for data standardization, effective model transfer across regions, and better methods to leverage unlabeled data. Evaluation and mitigation of algorithmic biases to improve transparency and trust in AI applied to agriculture.	[13–24]
	Socio-environmental dimensions of innovation	Recognition of inequalities in access to advanced technologies and the lack of studies on long-term environmental impacts. A proposal to integrate social, economic, and environmental indicators to guide balanced technological development.	[13,14,15,16,12,4,1,14, [3–17]]
	Integrated multidisciplinary approaches	Recommendation for holistic models combining hydrological, agricultural, and socioeconomic data to capture systemic complexity. Promotion of collaborative pilot projects and ethical protocols for responsible and equitable technology deployment.	[1, 4, 7, 8, 12–20]

Table 1.
Comparative matrix with the main contributions Comparative matrix of different research articles by category, main topic, and their main contributions..

and digital transformation. It reveals a global shift toward data-driven, technology-enabled environmental and agricultural management, while highlighting persistent challenges related to equity, scalability, and systemic integration.

Further, the integration of socioeconomic factors with biophysical data remains underdeveloped, resulting in models that insufficiently account for human dimension dynamics, such as market access, farmer behavior, and policy influences. Smallholder farmers often remain marginalized from technological benefits due to insufficient extension services and fragmented delivery mechanisms [6].

3.6 Comparative analysis matrix

The matrix synthesizes a broad and interdisciplinary body of knowledge focused on the intersection of water sustainability, climate change, agricultural innovation, and digital transformation. It reveals a global shift toward data-driven, technology-enabled environmental and agricultural management, while also highlighting persistent challenges related to equity, scalability, and systemic integration.

3.6.1 The global hydrological cycle under pressure: From depletion to digital monitoring

A central theme in the matrix is the growing instability of the global hydrological cycle, increasingly disrupted by climate change and anthropogenic pressures. One of the most critical issues identified is the rapid depletion of groundwater resources, particularly in densely populated and agriculturally intensive regions such as India and China. Satellite data from the GRACE (Gravity Recovery and Climate Experiment) mission have been instrumental in quantifying these losses, revealing alarming rates of aquifer decline primarily driven by unsustainable agricultural irrigation [1, 4, 11–16].

These satellite observations are not standalone solutions but are most effective when integrated with in-situ measurements and hydrological models, enhancing the accuracy and spatial resolution of regional assessments. This hybrid approach enables more reliable forecasting and supports better-informed water governance.

Furthermore, the concept of non-stationarity in the terrestrial hydrological cycle is emphasized – meaning that historical patterns of rainfall, runoff, and recharge can no longer be assumed stable. This paradigm shift challenges traditional water infrastructure planning and early warning systems, which often rely on past climatic behavior. As a result, adaptive, dynamic models are urgently needed.

To address these challenges, the matrix highlights the integration of remote sensing and AI as a transformative strategy. By combining satellite-derived data on soil moisture, evapotranspiration, and vegetation health with ML models, researchers and policymakers can monitor drought trends and agricultural water consumption in near real-time. Applications in regions such as California demonstrate how such systems can optimize irrigation scheduling and support drought resilience.

3.6.2 The digital transformation of agriculture: Precision, intelligence, and inclusion

The matrix underscores a technological revolution in agriculture, driven by AI, the IoT, and advanced remote sensing. Precision agriculture and smart greenhouse technologies are shown to significantly enhance productivity, resource efficiency, and crop yields. For example, IoT sensors combined with AI-driven climate control in greenhouses allow for real-time adjustments in temperature, humidity, and nutrient delivery [30].

However, the analysis also identifies significant barriers to adoption, especially among smallholder farmers. High initial investment costs, lack of rural internet connectivity, and limited technical training create a digital divide that risks excluding vulnerable populations from the benefits of innovation. Therefore, the matrix stresses the importance of targeted training programs, subsidies, and extension services to ensure equitable access and widespread adoption.

In terms of technical innovation, the use of deep learning architectures with self-attention mechanisms is proposed to improve crop classification and water use estimation from satellite imagery – even under challenging conditions such as cloud cover or spectral noise. These models benefit from self-supervised learning, which leverages vast amounts of unlabeled data, reducing the dependency on expensive, manually annotated datasets. This is particularly valuable in regions with limited monitoring infrastructure.

3.6.3 Land use change and urban expansion: Competing pressures on the Earth's surface

Urbanization is identified as a major driver of land-use change, with urban areas expanding faster than population growth. This trend leads to the conversion of fertile agricultural land into built environments, threatening food security and increasing pressure on surrounding ecosystems [5].

Despite these negative impacts, the matrix notes positive countertrends, such as reforestation and improved land management practices in certain regions, which help mitigate land degradation and enhance carbon sequestration. To monitor these dynamics effectively, the integration of multi-sensor, multi-temporal, and multi-spectral satellite data is essential.

Advanced classification methods, particularly those combining deep neural networks with data fusion techniques, have been shown to significantly improve the accuracy of LULC mapping. These tools overcome common limitations, such as cloud interference and spatial heterogeneity, enabling more reliable detection of deforestation, urban sprawl, and agricultural encroachment [20].

3.6.4 Climate change impacts on agriculture and the WEFC nexus

Climate change is already affecting agricultural productivity, with evidence of declining crop yields due to rising temperatures, shifting precipitation patterns, and an increased frequency of extreme weather events [21]. These impacts are not isolated but are deeply interconnected with other systems, leading to the emergence of the WEFC nexus as a key analytical framework [7].

This holistic perspective reveals critical trade-offs and synergies. For instance:

- Irrigation increases water demand and energy use (for pumping) but can also provide a cooling effect in agricultural landscapes, mitigating local heat stress.

- Fertilizer use boosts food production but contributes to nitrous oxide emissions, a potent GHG.

- Agricultural expansion often leads to deforestation, reducing carbon sinks and increasing emissions.

- The matrix emphasizes that sustainable policies must account for these interdependencies. AI-based predictive models are highlighted as essential tools for anticipating climate extremes and integrating climate risk into agricultural planning and public policy, promoting resilience and social inclusion.

3.6.5 ML and drought forecasting: Toward proactive water management

Drought monitoring has evolved from reactive to predictive, thanks to advances in machine learning and multi-source data integration. The development of composite drought indices, which combine temperature, soil moisture, and precipitation data using methods such as principal component analysis and weighted averages, allows for more accurate and timely assessments across diverse regions [1, 21]

Deep learning models, particularly LSTM (Long Short-Term Memory) networks with attention mechanisms, have proven effective in forecasting drought conditions by capturing temporal dependencies in climate data. These models are validated against meteorological and agricultural indicators and are especially useful in data-scarce or ungauged basins, where traditional monitoring is limited [31].

By incorporating long-range climate forecasts into machine learning frameworks, it becomes possible to anticipate droughts months in advance, enabling early interventions such as water rationing, crop switching, or emergency support for farmers.

3.6.6 Ethical, security, and socioeconomic dimensions of technological adoption

Beyond technical capabilities, the matrix raises critical questions about ethics, data security, and equity in the deployment of digital agricultural systems. AI platforms, while powerful, can introduce vulnerabilities related to data privacy, algorithmic bias, and a lack of transparency.

To address these concerns, the matrix proposes the use of blockchain technology to enhance data traceability, ownership, and trust. Blockchain can ensure that farmers retain control over their data and that supply chains are transparent and verifiable – key elements for sustainable certification and fair trade.

Moreover, the analysis highlights persistent inequalities in access to technology, which could exacerbate existing social and economic disparities. Therefore, it calls for inclusive regulatory frameworks, ethical guidelines, and participatory design to ensure that technological advancements benefit all stakeholders, especially marginalized groups [29].

3.6.7 Capacity building, policy, and international cooperation

The successful adoption of smart agricultural technologies depends not only on innovation but also on human capital and supportive policies. The matrix identifies financial, infrastructural, and educational barriers as major constraints, particularly in rural and developing regions [14].

To overcome these, it recommends:

- Thematic training programs tailored to local needs.
- Government subsidies to lower the entry cost of smart technologies.
- Advisory services to guide farmers in technology use and decision-making.
- Additionally, international pilot projects in climate-smart agriculture are presented as valuable platforms for knowledge exchange, comparative learning, and technological standardization. These collaborative efforts can accelerate innovation and adaptation across different agroecological zones.

At the policy level, the matrix critiques current national strategies for digital transformation in agriculture, calling for more inclusive, environmentally assessed, and circular economy-aligned frameworks. Such policies would not only promote technological adoption but also ensure long-term sustainability and social impact.

3.6.8 Future directions and research gaps

The matrix concludes with a forward-looking perspective on unresolved challenges and emerging opportunities [24]:

- Data standardization and model transferability remain major hurdles. Models trained in one region often fail in others due to differences in climate, soil, or cropping systems.
- There is a need for better methods to leverage unlabeled data through semi-supervised and self-supervised learning, reducing reliance on costly labeled datasets.
- Algorithmic bias must be systematically evaluated and mitigated to ensure fairness and transparency in AI applications.
- Long-term socio-environmental impacts of digital agriculture are still poorly understood, with few studies assessing ecological trade-offs or social equity outcomes.
- To address these gaps, the matrix advocates for integrated, multidisciplinary approaches that combine hydrological, agricultural, climatic, and socioeconomic data. It also promotes collaborative pilot projects and ethical deployment protocols to ensure that technological progress is both responsible and equitable.

3.7 Detailed analysis and synthesis of key issues

Table 2 summarizes the deficiencies identified, their causes, and potential solutions or research needs identified in the various articles analyzed.

Domain	Key gap	Cause(s)	Potential solutions/research needs
Sustainable agriculture	System fragmentation and policy disconnects	Lack of holistic system integration; separation of urban-rural studies	Develop integrated system frameworks, considering feedback and trade-offs.
Sustainable agriculture	Under-addressed biodiversity–ecosystem service interactions	Segregated land use models, scalability challenges	Enhance ecoagriculture approaches, address governance and financial barriers.
Sustainable agriculture	Insufficient socioeconomic and cultural factors	Focus on biophysical factors; lack of social inclusion	Incorporate socio-cultural research; explore HRM flexibility in agriculture.
Sustainable agriculture	Climate resilience and abiotic stress management	Emerging technologies are not fully integrated or accessible	Combine genetics, bioinformatics, and AI phenotyping for stress tolerance.
Sustainable agriculture	Validation and standardization in risk assessment	Static indices, poor hazard definition, lack of scenario testing	Advanced dynamic modeling, validation protocols, and integrated adaptive strategies.
Remote sensing	Scale and resolution limitations	Local focus, coarse global data, dependency on ground truth	Develop multi-scale approaches, leverage crowdsourcing, and hybrid data fusion.
Remote sensing	Limited sensor and data fusion	Use of single sensors, restricted spectral bands	Integrate multi-sensor datasets; emphasize platforms such as Sentinel-2.
Remote sensing	ML model challenges	Black-box problems, inconsistent accuracy, and lack of standardized data	Enhance interpretability, create open benchmarks, incorporate diverse crops/environments.
Remote sensing	Implementation barriers for real-time decision support	Data gaps, computational complexity, user accessibility issues	Develop user-friendly platforms, improve data assimilation, and foster stakeholder engagement.
Remote sensing	Weak multidisciplinary integration	Data silos, disciplinary mismatches	Promote interdisciplinary collaboration, standardize data formats, and techniques.

Source: own elaboration.

Table 2.
Summary matrix of identified gaps.

3.7.1 Systemic fragmentation and policy gaps

Studies examining the relationship between urban sprawl and food security reveal a critical failure in incorporating the complex, interlinked nature of urban–rural food systems [32]. Policies too often address segmented parts of the system, neglecting feedback loops and dynamics essential for sustainable transformation. This shortcoming extends to the design of forest conservation programs and agricultural policies, where moderate effectiveness is often tied to implementation in low-pressure environments, lacking alignment with local–global goals [33].

3.7.2 Biodiversity conservation in agricultural landscapes

The dominant model of separating conservation from agricultural production is being reassessed as untenable due to increased agricultural ecological footprints [34].

Ecoagriculture landscapes have been proposed, incorporating native biodiversity alongside farming activities. Yet, challenges in making these approaches financially and institutionally feasible remain considerable, necessitating coordinated efforts across scientists, practitioners, and policymakers, with an emphasis on enabling governance frameworks.

3.7.3 Integrating socioeconomic and human factors

The role of flexible work arrangements and evolving human resource management in agriculture reflects an emerging area seldom addressed in conjunction with sustainability goals. Evidence from non-agricultural sectors shows that flexibility and job satisfaction have positive links with performance and renewal capacity [35]. Incorporating such perspectives into agricultural labor management can strengthen community resilience and sustainability.

3.7.4 Advances and limitations in climate-resilient crop production

Cutting-edge biotechnologies, including CRISPR gene editing, marker-assisted selection, and omics technologies, have been increasingly applied to enhance drought resistance and mitigate biotic stresses in staple crops [36]. Precision agriculture tools, such as IoT devices and remote sensing platforms, further optimize water use and resource efficiency in arid environments. Nonetheless, the interdisciplinary integration of these advances requires expanded field applicability, affordability, and knowledge transfer mechanisms.

3.7.5 Remote sensing: Data, algorithms, and applications

The Sentinel-2 satellite constellation, with its twin-platform, improved spectral and spatial resolution, exemplifies advances in agricultural monitoring, facilitating the detection of abiotic and biotic stresses [37]. However, limitations persist in standalone remote sensing approaches, necessitating fusion with ground-truth and complementary techniques to enhance accuracy and operational relevance.

Machine learning applications such as CNN, LSTM, and Random Forest algorithms have achieved notable success in crop classification and yield prediction using time-series satellite data [38]. Yet, issues of model transparency, generalizability across diverse agro-ecological zones, and consistency impede broader trust and adoption.

Moreover, water-use efficiency improvements via remote sensing-enabled irrigation scheduling are promising but require better data assimilation systems and user interface designs to transition from research to farmer application [39].

3.8 Future directions and research priorities

Addressing these gaps demands multidimensional research and coordinated innovation. Priority areas include the following:

- Designing integrated agricultural systems frameworks, incorporating socioeconomic, ecological, and technological dimensions, to overcome policy fragmentation.

- Scaling ecoagriculture landscape approaches with attention to governance, financing, and local stakeholder involvement.
- Developing and validating dynamic, scenario-based risk assessment models for climate and drought vulnerabilities.
- Advancing multi-sensor remote sensing data fusion and leveraging AI models with improved interpretability and robustness.
- Enhancing user-oriented, real-time decision-support platforms that bridge the gap between big data analytics and practical agricultural management.
- Promoting transdisciplinary collaborations and data standards to effectively integrate agronomic, ecological, socioeconomic, and remote sensing knowledge bases.

By systematically identifying these gaps and synthesizing insights across disciplines, this review highlights critical bottlenecks within sustainable agriculture and remote sensing methodologies. Bridging these divides through interdisciplinary strategies and embracing technological advancements will be pivotal for resilient, sustainable agricultural systems capable of meeting global challenges.

4. Conclusion(s)

The integration of sustainable agriculture practices with climate change adaptation and mitigation strategies is critical for ensuring global food security and environmental sustainability. Remote sensing technologies, coupled with machine learning and AI, offer unprecedented capabilities for monitoring agricultural systems, informing precision farming, and quantifying environmental effects such as GHG emissions. These technologies enable more precise and timely data acquisition at multiple scales, facilitating informed decision-making and policy development.

However, significant limitations persist, notably the uneven adoption of these technologies in developing regions, the challenge of addressing smallholder farmers' needs, and the insufficient incorporation of socioeconomic and policy dimensions into technical models. Future research should aim to develop scalable, integrative frameworks that merge biophysical data with socioeconomic variables to foster inclusive, climate-resilient agricultural systems. Additionally, enhancing data accessibility, refining AI algorithms for localized contexts, and fostering multi-stakeholder collaborations will be vital to overcoming current barriers. Addressing these gaps will ensure that remote sensing and AI technologies fulfill their promise in advancing sustainable agriculture amid ongoing climate change.

Conflict of Interest

The author declares no conflicts of interest.

Author details

Jhon Jairo Castañeda Fandiño
University of Cundinamarca, Bogotá, Colombia

*Address all correspondence to: jjairocastaneda@ucundinamarca.edu.co

References

[1] Abbas T, Shoaib M, Albano R, Baig MAI, Ali I, Farid HU, Ali MU. Artificial-Intelligence-based investigation on Land Use and Land Cover (LULC) changes in response to population growth in South Punjab, Pakistan. Land. 2025;**14**:154. DOI: 10.3390/land14010154

[2] Changxin Y, Yuan Z, Chen X, Zhong R, Huang L. Revealing climate-induced patterns in crop yields and the water-energy-food-carbon nexus: Insights from the pearl river basin. Water. 2024. DOI: 10.3390/w16243693

[3] Wani SUHASP. New paradigm for transforming Indian agriculture to climate-resilient and sustainable agriculture is a must. Journal of Agrometeorology. 2023. DOI: 10.54386/jam.v25i1.2011

[4] Sellam V, Kannan N, Pandi SS, Manju I. Enhancing sustainable agriculture using attention convolutional bidirectional Gated recurrent based modified leaf in wind algorithm: Integrating AI and IoT for efficient farming. Sustainable Computing: Informatics and Systems. 2025. DOI: 10.1016/j.suscom.2025.101160

[5] Zeng H, Xie Y, Elnashar A, Wang S, Zhao H, Junbin L, Bingfang W. Increasing Cropland Area and Its Associated Human-Induced Water Consumption Put Ebinur Lake at Risk of Drying Up. Journal of Remote Sensing. 2025. DOI: 10.34133/remotesensing.0496

[6] Munaganuri RK, Rao YN. 2023. "PAMICRM: Improving precision agriculture through multimodal image analysis for crop water requirement estimation using multidomain remote sensing data samples". IEEE Access. DOI: 10.1109/access.2024.3386552

[7] Uyar N, Uyar A. Assessing climate change impacts on cropland and greenhouse gas emissions using remote sensing and machine learning. Atmosphere. 2025;**16**:418. DOI: 10.3390/atmos16040418

[8] Ye C, Yuan Z, Chen X, Zhong R, Huang L. Revealing climate-induced patterns in crop yields and the water-energy-food-carbon nexus: Insights from the pearl river basin. Water. 2024;**16**:3693. DOI: 10.3390/w16243693

[9] Veerakachen MRAW. Farmdee-Mesook: An Intuitive GHG Awareness Smart Agriculture Platform; 2024. DOI: 10.3390/agronomy15081772

[10] Atashov <>, Abbasov <>, Gurbanzade <>. Government policy on ensuring food security: Bibliometric analysis. Problems and Perspectives in Management. 2025;**23**:496–514. DOI: 10.21511/ppm.23(1).2025.38

[11] Nie W, et al. Nonstationarity in the global terrestrial water cycle and its interlinkages in the Anthropocene. Proceedings of the National Academy of Sciences of the United States of America. 2024;**121**(45). DOI: 10.1073/pnas.2403707121

[12] Tang P, et al. Deep learning with multi-scale temporal hybrid structure for robust crop mapping. ISPRS Journal of Photogrammetry and Remote Sensing. 2024;**209**:117–132

[13] Nagal B, Prabhakar AK, Pal M. Groundwater storage analysis using geospatial techniques: A comprehensive review over Asian Region. Journal of Engineering Science and Technology Review. 2025;**18**(1):187–198. DOI: 10.25103/jestr.181.19

[14] Beisekenov N, Banakinaou W, Ajayi AD, Hasegawa H, Tadao A. Remote sensing-based soil organic carbon monitoring using advanced machine learning techniques under conservation agriculture systems. Smart Agricultural Technology. 2025;**11**. DOI: 10.1016/j.atech.2025.101036

[15] Munaganuri RK, Rao YN. PAMICRM: Improving precision agriculture through multimodal image analysis for crop water requirement estimation using multidomain remote sensing data samples. IEEE Access. 2024;**12**:52815–52836. DOI: 10.1109/ACCESS.2024.3386552

[16] Wang H, et al. Improving the crop classification performance by unlabeled remote sensing data. Expert Systems with Applications. 2024;**236**. DOI:10.1016/j.eswa.2023.121283

[17] Albarakati HM, et al. A novel deep learning architecture for agriculture land cover and land use classification from remote sensing images based on network-level fusion of self-attention architecture. IEEE Journal of Selected Topics in Applied Earth Observations and Remote Sensing. 2024;**17**:6338–6353

[18] Chen B, Huang B, Xu B. Multi-source remotely sensed data fusion for improving land cover classification. ISPRS Journal of Photogrammetry and Remote Sensing. 2017;**124**:27–39. DOI: 10.1016/j.isprsjprs.2016.12.008

[19] Zhang J, Mu Q, Huang J. Assessing the remotely sensed Drought Severity Index for agricultural drought monitoring and impact analysis in North China. Ecological Indicators. 2016;**63**:296–309. DOI: 10.1016/j.ecolind.2015.11.062

[20] Al-Raeei M. Artificial intelligence for climate resilience: Advancing sustainable goals in SDGs 11 and 13 and its relationship to pandemics. Discover Sustainability. 2024;**5**(1). DOI: 10.1007/s43621-024-00775-5

[21] Zhou J, Fan Y, Guan Q, Feng G. Research on drought monitoring based on deep learning: A case study of the Huang-Huai-Hai Region in China. Land (Basel). 2024;**13**(5). DOI: 10.3390/land13050615

[22] Li L, Wang S, Bo Y, Yang B, Li X, Liu K. Spatial-temporal evolution and cooling effect of irrigated cropland in inner Mongolia Region. Remote Sensing (Basel). 2024;**16**(24). DOI: 10.3390/rs16244797

[23] Mokhtar A, He H, Nabil M, Kouadri S, Salem A, Elbeltagi A. Securing China's rice harvest: Unveiling dominant factors in production using multi-source data and hybrid machine learning models. Scientific Reports. 2024;**14**(1). DOI: 10.1038/s41598-024-64269-0

[24] Triantakonstantis D, Karakostas A. Soil organic carbon monitoring and modelling via machine learning methods using soil and remote sensing data. Agriculture (Switzerland). 2025;**15**(9). DOI: 10.3390/agriculture15090910

[25] Karahan H, Cetin M, Can ME, Alsenjar O. Developing a new ANN model to estimate daily actual evapotranspiration using limited

climatic data and remote sensing techniques for sustainable water management. Sustainability (Switzerland). 2024;**16**(6). DOI: 10.339 0/su16062481

[26] Jiao W, Tian C, Chang Q, Novick KA, Wang L. A new multi-sensor integrated index for drought monitoring. Agricultural and Forest Meteorology. 2019;**268**:74–85. DOI: 10.1016/j.agrformet.2019.01.008

[27] Shen R, Huang A, Li B, Guo J. Construction of a drought monitoring model using deep learning based on multi-source remote sensing data. International Journal of Applied Earth Observation and Geoinformation. 2019;**79**:48–57. DOI: 10.1016/j. jag.2019.03.006

[28] Rhee J, Im J. Meteorological drought forecasting for ungauged areas based on machine learning: Using long-range climate forecast and remote sensing data. Agricultural and Forest Meteorology. 2017;**237-238**:105–122. DOI: 10.1016/j.agrformet.2017.02.011

[29] Raksapatcharawong M, Veerakachen W. Farmdee-Mesook: An intuitive GHG awareness smart agriculture platform. Agronomy. 2025;**15**(8):1772. DOI: 10.3390/ agronomy15081772

[30] Ramadhani F, Pullanagari R, Kereszturi G, Procter J. Mapping a cloud-free rice growth stages using the integration of PROBA-v and sentinel-1 and its temporal correlation with sub-district statistics. Remote Sensing (Basel). 2021;**13**(8). DOI: 10.3390/ rs13081498

[31] Wang L, Wang J, Liu Z, Zhu J, Qin F. Evaluation of a deep-learning model for multispectral remote sensing of land use and crop classification. The Crop Journal. 2022;**10**(5):1435–1451. DOI: 10.1016/j.cj.2022.01.009

[32] Hatab AA, Cavinato MER, Lindemer A, Lagerkvist CJ. Urban sprawl, food security and agricultural systems in developing countries: A systematic review of the literature. Elsevier BV. 2019. DOI: 10.1016/j. cities.2019.06.001

[33] Brner J, Schulz D, Wunder S, Pfaff A. The effectiveness of forest conservation policies and programs. Annual Reviews. 2020. DOI: 10.1146/a nnurev–resource–110119–025703

[34] Scherr S, McNeely JA. Biodiversity conservation and agricultural sustainability: Towards a new paradigm of ecoagriculture landscapes. Royal Society. 2007. DOI: 10.1098/ rstb.2007.2165

[35] Davidescu AA, Apostu SA, Paul A, Cuneanu I. Work flexibility, job satisfaction, and job performance among romanian employees implications for sustainable human resource management. Multidisciplinary Digital Publishing Institute. 2020. DOI: 10.3390/ su12156086

[36] Xing Y, Wang X. Precision agriculture and water conservation strategies for sustainable crop production in arid regions. Multidisciplinary Digital Publishing Institute. 2024;10.3390/plants13223184

[37] Segarra J, Buchaillot ML, Araus JL, Kefauver SC. Remote sensing for precision agriculture: Sentinel-2 improved features and applications. Multidisciplinary Digital Publishing Institute. 2020;10.3390/agronomy 10050641

[38] Muruganantham P, Wibowo S, Grandhi S, Samrat NH, Islam N. A systematic literature review on crop yield prediction with deep learning and remote sensing. Multidisciplinary Digital Publishing Institute. 2022. DOI: 10.3390/rs14091990

[39] Zhao H, Di L, Sun Z. WaterSmart-GIS: A web application of a data assimilation model to support irrigation research and decision making. Multidisciplinary Digital Publishing Institute. 2022. DOI: 10.3390/ijgi11050271

Chapter 2

Beyond Proximity: Unpacking the Drivers of Diverse Urban Park Visitation

Jianqiang Huang and Patrick O'Connor

Abstract

Access to natural environments is crucial for public health in urban areas, yet the specific park qualities that attract visitors remain underexplored. While factors like proximity are known to influence park use, the role of biodiversity is less clear. This chapter investigates the relationship between residential location and park visitation in South Australia. Using survey data from 2021 to 2023, we analysed the diversity of park choices between urban and peri-urban residents with the Herfindahl–Hirschman Index and explored the potential drivers of these patterns, including ecological and sociodemographic factors. Our results reveal that urban residents explore a significantly wider variety of parks compared to their peri-urban counterparts. However, the drivers of this behaviour are complex. Within the urban area, visitation was not directly correlated with standard biodiversity metrics. Instead, preferences were linked to specific landscape features, such as the presence of native vegetation over shrubland, alongside park accessibility and management. These findings suggest that creating attractive urban green spaces requires a nuanced approach. Investing in a portfolio of high-quality, accessible, and well-managed parks that feature preferred native ecosystems is a vital strategy for promoting public health and fostering a greater connection between city dwellers and nature.

Keywords: urban green infrastructure, park preference, nature-based solutions, urban ecology, urban planning, human well-being

1. Introduction

National parks and protected areas are cornerstones of global conservation efforts, preserving critical ecosystems, safeguarding biodiversity, and providing essential environmental services [1]. Beyond their ecological significance, these natural landscapes are increasingly recognised as vital resources for public health and well-being, particularly for the world's growing urban population. Access to green spaces offers city dwellers crucial opportunities for physical activity, which can mitigate the risks of chronic diseases, and for psychological restoration, helping to reduce stress and mental fatigue. These environments also serve as social hubs, fostering community connection and a shared sense of place. As urbanisation continues to expand, the parks and natural areas at the urban fringe become ever more important as accessible havens for nature-based recreation and health promotion.

Considerable research has been dedicated to understanding what drives people to visit parks [2]. The literature consistently shows that logistical factors like proximity and accessibility are primary determinants of park use. People are more likely to visit parks that are closer to their homes, easy to get to via public or private transport, and equipped with amenities that support their recreational goals. Consequently, much of the discourse around equity in park access has focused on the spatial distribution of green spaces, aiming to ensure that all communities have a park within a reasonable distance. While this focus is undeniably important, it often treats green space as a uniform good, implicitly assuming that one park is as good as another if it is accessible.

However, this perspective overlooks the intrinsic qualities of the parks themselves, particularly their ecological richness. A critical and underexplored question is whether the biodiversity of a park influences visitation patterns. Is a park with a complex, thriving ecosystem more attractive to visitors than a more manicured or ecologically simple one? The existing literature offers little direct empirical evidence on this matter. While studies confirm the effects of proximity and accessibility, the role of biodiversity as a potential driver of visitation remains largely unassessed. This represents a significant gap in our understanding of human-nature interactions, as it is plausible that the very 'naturalness' that conservation efforts seek to protect is a key part of what draws people to these spaces in the first place.

Furthermore, the connection between biodiversity and the health benefits derived from park visits is often assumed rather than demonstrated. The potential for biodiverse environments to offer superior psychological benefits is a compelling hypothesis, for instance, through a greater fascination and a stronger sense of being away from daily life. However, most of these claims lack robust empirical support in the context of park visitation. If richer natural settings do indeed lead to more restorative experiences or encourage longer and more frequent visits, then biodiversity is not just a conservation goal but also a public health asset. Failing to investigate this link means we may be underestimating the full value of our most ecologically significant protected areas.

This chapter addresses these research gaps by providing a direct empirical analysis of the interplay between residential location, park visitation patterns, and biodiversity. We investigated two primary questions: first, do the park visitation patterns of urban residents differ from those of their peri-urban counterparts in terms of the number and variety of parks they visit? Second, is there a demonstrable link between these visitation patterns and the biodiversity of the parks they choose to visit? By examining these questions, we aim to move beyond accessibility alone and provide new insights into the role of biodiversity as a driver of human engagement with nature. Ultimately, this work seeks to inform more integrated strategies for urban planning, public health, and conservation that recognise the profound connections between healthy ecosystems and healthy communities.

2. Methods

2.1 Parks visitation survey

We utilised data from McGregor Tan's South Australian Parks Visitation Survey. The survey, which has been running since 2015, collects responses about which parks people have visited and the total number of visits during the 12 months prior to the survey. This study uses data from 2021 to 2023 to focus on recent trends. To align with our study's scope, only responses about parks within the respondents' own Natural Resource Management (NRM) region were included.

2.2 Herfindahl–Hirschman Index

The HHI is a common measure of market concentration. An HHI below 1,500 indicates low concentration, a score between 1,500 and 2,500 indicates moderate concentration, while a score above 2,500 indicates high concentration. In this study, we use the HHI to measure the concentration of park visits. A lower HHI suggests that visits are distributed more evenly across a wider range of parks, indicating more diverse visitation patterns.

The original HHI is calculated by summing the squares of the market shares of each firm. Here, it is the sum of squared shares for each park:

$$HHI = \sum_{i=1}^{N} \left(\frac{Park_{sharei}}{Total_{share}} \right)^2$$

where *Park Share*$_i$ refers to the proportion of people who report at least one visit to a specific $park_i$. Since participants are allowed to choose multiple parks, the sum of shares within an NRM could exceed 100%. To address this, shares were divided by the total share in each NRM for normalisation.

To compare urban and peri-urban areas, a weighted HHI for all peri-urban NRMs (NRM 2–8) was computed by weighting the HHI of each NRM by its volume of survey responses:

$$HHI_{peri-urban} = \frac{\sum_{i=2}^{8} (HHI_i \times Respondent_i)}{\sum_{i=2}^{8} Respondent_i}$$

2.3 Biodiversity index

Biodiversity data for each park were obtained from the South Australian Department for Environment and Water. Specifically, we used land cover data (2010–2015) to calculate the proportion of native vegetation, statewide vegetation data to classify major vegetation groups (MVG) and subgroups (MVS), and species lists from NatureMaps to assess flora and fauna diversity.

Two widely used biodiversity indices were employed: the Shannon–Wiener Index (H′) and Simpson's Diversity Index (1 - D). The Shannon–Wiener Index measures species diversity by considering both richness (the number of species) and evenness (how evenly individuals are distributed among species). Simpson's Diversity Index measures the probability that two randomly selected individuals will be from different species. For both indices, a higher score indicates greater biodiversity.

2.4 Negative binomial regression

The number of visits in the survey is a count variable, collected through the question, 'What is the total number of visits you have made in the last 12 months?'

They are highly skewed, with many zero or low values and a long tail of high values. If we want to draw a model for visit numbers, a standard OLS linear regression is inappropriate in this case because it assumes a continuous, normally distributed dependent variable. While Negative Binomial Regression, a generalisation of the Poisson model, is capable of accounting for over-dispersion and only predicts non-negative integers, it is a good fit for the visit number model.

In this chapter, we use glm.nb function from the MASS package in R for statistical analysis.

3. The diverse choices of park visits in the Adelaide metro area

During the survey, participants were asked which national parks they had visited in the past 12 months. While South Australia has over 300 national parks, only 10–15 parks in their local NRM were listed. The rest fell under 'Other – Please specify'. Consequently, some parks were mentioned by only a few respondents and had extremely low visitation counts, making it difficult to draw meaningful inferences. We excluded these parks to allow for a more reliable comparison. From the responses, we collected data on the parks visited, the total number of annual visits, barriers to visitation, and demographic status.

From this visitation data, we identified the most and least visited parks in each NRM (**Table 1**). We observed some uneven visitation patterns in certain NRMs. For

	Number of parks counted[b]	Most visited parks[c]	Visit rate (%)[d]	Least visited parks[c]	Visit rate (%)[d]
NRM1[a]	16 out of 56	Belair (NP)	34.7	Black Hill (CP)	7.6
		Cleland (CP), including the Waterfall Gully to Mount Lofty trail.	27.5	Scott Creek (CP)	4.4
		Morialta (CP)	22.4	Para Wirra (RP)	0.4
NRM2	11 out of 55	Murray River (NP)	42.0	Ngarkat (CP)	5.1
		Monarto (CP)	31.8	Mount Magnificent (CP)	4.5
		Coorong (NP)	27.4	Morgan (CP)	1.3
NRM3	15 out of 26	Flinders Chase (NP)	80.0	Baudin (CP)	3.3
		Vivonne Bay (CP)	60.0	Cape Willoughby (CP)	3.3
		Pelican Lagoon (CP)	50.0	Dudley (CP)	3.3
				Lashmar (CP)	3.3
				Lesueur (CP)	3.3
				Simpson (CP)	3.3
				The Pages (CP)	3.3

	Number of parks counted[b]	Most visited parks[c]	Visit rate (%)[d]	Least visited parks[c]	Visit rate (%)[d]
NRM4	8 out of 66	Beachport (CP)	44.3	Guichen Bay (CP)	8.5
		Piccaninnie Ponds (CP)	34.9	Desert Camp (CP)	2.8
		Canunda (NP)	28.3	Ewens Ponds (CP)	1.9
NRM5	9 out of 80	Lincoln (NP)	49.0	Fowlers Bay (CP)	2.9
		Coffin Bay (NP)	45.2	Neptune Islands (CP)	2.9
		Franklin Harbor (CP)	14.4	Whyalla (CP)	1.9
NRM6	7 out of 22	Mount Remarkable (NP)	35.0	Mount Brown (CP)	4.9
		Innes (NP)	25.2	Clinton (CP)	2.9
		Telowie Gorge (CP)	19.4	Troubridge Island (CP)	2.9
NRM7	8 out of 8	Great Australian Bight Marine (NP)	45.0	Pureba (CP)	15.0
		Nullarbor (NP)	45.0	Tallaringa (CP)	15.0
		Yumbarra (CP)	30.0	Wahgunyah (CP)	10.0
				Mamungari (CP)	5.0
NRM8	3 out of 19	Flinders Ranges (NP)	59.1	Lake Gairdner (NP)	4.6

Note:
a) Names of each NRM are Adelaide and Mt Lofty Ranges, South Australian Murray-Darling Basin, Kangaroo Island, South East, Eyre Peninsula, Northern & Yorke, Alinytjara Wilurara, and South Australian Arid Lands, respectively.
b) Number of parks visited out of the total number in that NRM. The total park number is derived from the McGregor Tan survey. It's not necessarily the same as the full national park list, but it should somehow represent most of them.
c) After eliminating parks with very few reported visits, we selected three parks with the most visits and the least visits. The number may exceed three if parks shared the same visitation rate. NRM 8 had only three parks visited; thus, only one park was picked each of the most visited and the least visited. NP: National Parks; CP: Conservation Parks; RP: Recreation Parks.
d) Visit rate was calculated within each NRM.

Table 1.
Most and least visited parks in each NRM.

example, in NRM 3 and 5, a dominant park emerged, and the visit rate dropped sharply after the three most visited parks. We could also see that several parks had no visit records during the three-year survey in NRM 4 and 5.

We grouped NRMs by urban (NRM 1, Adelaide metro area) and peri-urban (NRM 2–8) regions and applied the HHI tool to dive deeper into park visitation (**Table 2**). The HHI for the urban region is 888, indicating low concentration and a diverse range of parks visited. In contrast, the weighted HHI for peri-urban areas is 2,189, signifying a moderate-to-high concentration of visits to fewer parks.

We could see the different distances to parks in different NRMs, which implies that people in the Adelaide metro area can access parks more easily. However,

		NRM1	NRM2	NRM3	NRM4	NRM5	NRM6	NRM7	NRM8	2–8
HHI		888	1,523	1,382	1982	2,540	2,404	1,731	6,800	2189
Median visits		4	4	5	4	4	3	5	4	4
Median distance (km)[a]		44	140	220	427	621	268	1,160	423	417
Visitor (%)[b]		76	78	97[e]	69	66	62	75[e]	77[e]	72
Barriers[c]	Time	4	4	4	4	4	4	4	4	4
	Safety	1	1	1	1	1	1	1	1	1
	Facilities	2	2	2	2	2	2	2	1	2
	Not thinking	2	2	1	2	2	2	2	2	2
	Entry cost	2	2	2.5	2	2	2	1	2	2
	Transportations	2	2	1	2	2	1	1.5	1	2
Shannon–Wiener Index (H′)[d]		5.98	4.88	5.41	2.30	4.97	5.98	0.76	-	-
Simpson's Diversity (1-D)[d]		0.99	0.98	0.98	0.87	0.99	1.00	0.31	-	-

Note:
a) Median distance represents the median of individual mean distance from participant postcode to the top-visited national parks within each NRM region. The original distance is the driving distance from the postcode area centroid to the park, derived from the Google Maps Platform.
b) Visitor percentage: participants who reported at least one visit divided by the total participants in each NRM.
c) Barriers were a median value of the 1–5 Likert scale. While not a numerical variable, it comes with an ordinal ranking of agreement (to what degree you agree with this statement). Thus, a median was used to identify the central tendency of barriers.
d) Biodiversity index was calculated in the most visited park in each NRM, which are Belair, Murray River, Flinders Chase, Beachport, Lincoln, Mount Remarkable, Great Australian Bight Marine, and Flinders Ranges, respectively.
e) NRM 3, 7, and 8 have less than 30 participants.

Table 2.
Brief information for each NRM.

research argues that remoteness doesn't increase visit rates in Australasia [3]. The survey also revealed that reported barriers to visitation were similar across both urban and peri-urban areas, with no difference in transportation barriers. What's more, the visitor percentage was similar, the median visit frequency was consistent, and the number of available parks was not significantly higher. Here comes the mysterious point: what led to more diverse park visits in the Adelaide metro area?

4. A closer look at the most and least visited parks

The preceding analysis revealed that urban residents have more diverse visitation patterns compared to their peri-urban counterparts, who tend to focus on a smaller group of popular parks. This raises a key question: what factors might account for this difference? A brief look at the data in **Table 2** suggests a potential clue, showing that the biodiversity of top-visited parks in the urban area is generally higher than that of other parks. This aligns with existing research [4, 5], which indicates that visitors may prefer parks with higher biodiversity.

To explore this further, we conducted a deeper analysis within the urban NRM 1. We compared the ecological characteristics of the three most-visited parks with those of the least-visited parks, examining land cover, vegetation groups, and biodiversity indices for both flora and fauna (**Table 3**).

		Top visited			Least visited		
		Belair	**Cleland**	**Morialta**	**Black Hill**	**Scott Creek**	**Para Wirra**
Land cover (%)[a]	1	99.14	99.72	99.71	97.98	98.79	96.48
	2	0.64	0.15	0.20	1.14	0.89	0.19
	3	0.35	0.08	0.20	0.85	0.08	0.11
Vegetation[b]	MVG	5	4, 5	5, 8, 21	5, 8, 16, 17, 21	4, 5	5, 8, 16, 21
	MVS	8, 9	4, 8, 9	8, 9, 26, 37	8, 9, 21, 26, 32, 37, 38	4, 8, 9	8, 9, 21, 26, 37, 38
Shannon–Wiener Index (H′)	Flora	6.12	5.94	5.40	5.22	6.13	5.41
	Fauna	3.72	3.99	4.07	4.22	3.75	4.03
Simpson's Diversity (1-D)	Flora	1.00	1.00	0.99	0.99	1.00	0.99
	Fauna	0.92	0.97	0.98	0.98	0.95	0.97

Note:
a)In Land Cover, 1 is native vegetation, 2 is non-native vegetation, and 3 is other.
b)List of main vegetation groups (G) and subgroups (S) is as follows:
Main vegetation group:
4. Eucalyptus Low Open Forest
5. Eucalyptus Woodland
8. Casuarina Forest and Woodland
16. Acacia Shrubland
17. Other Shrublands
21. Other Grassland, Herbland, Sedgeland, and Rushland
22. Chenopod Shrub, Samphire Shrub, and Forbland
Main vegetation subgroup
4. Eucalyptus forests with a shrubby understorey.
8. Eucalyptus woodlands with a shrubby understorey.
9. Eucalyptus woodlands with a grassy understorey
21. Other Acacia tall, open shrublands and shrublands
26. Casuarina and Allocasuarina forests and woodlands
32. Other shrublands
37. Other tussock grasslands
38. Wet tussock grassland, herbland, sedgeland, or rushland.

Table 3.
Most visited and least visited parks in NRM1.

The top-visited parks had slightly higher proportions of native vegetation and an absence of shrubland. This finding is consistent with a study of national forest visitors, which found a preference for sites with trees and shaded hiking trails compared to shrubland [6]. Potential reasons for this preference include safety concerns, as dense shrubs can block visibility and obstruct escape routes [7], or lower recreational satisfaction due to blocked views or difficulty spotting wildlife.

However, when we examined biodiversity, we found no significant difference between the most and least-visited parks. A possible explanation is that the national parks in the Adelaide metro area are well-managed and maintain a high level of biodiversity overall. Therefore, the preference for certain parks may be driven by factors like vegetation type or distance rather than biodiversity. Since urban residents have ample access to high-biodiversity parks, they may not need to travel further for such experiences. In peri-urban areas, however, where the biodiversity of top-visited parks is generally lower, the situation may be different.

5. What stops people from visiting national parks?

While the preceding analysis of ecological factors provides some insight, it does not fully explain the observed visitation patterns. To build a more comprehensive picture, we shifted our focus to sociodemographic drivers. Although the median number of visits was consistent across all eight NRMs, individual visitation habits varied significantly. This section presents a model analysis of visitation that accounts for demographics, perceived barriers, and fixed effects for time and region (**Table 4**).

	Estimate	Std.Error	p.value	IRR with CI
Intercept	13.380	0.287	0.000	46.499 (26.458, 85.458)***
Age (25–34)	-2.871	0.135	0.004	0.679 (0.513, 0.892)**
Age (35–44)	-2.578	0.132	0.010	0.712 (0.540, 0.931)**
Age (45–54)	-3.249	0.128	0.001	0.660 (0.503, 0.860)**
Age (55–64)	-4.021	0.128	0.000	0.598 (0.456, 0.779)***
Age (65–74)	-4.429	0.134	0.000	0.552 (0.416, 0.725)***
Age (75 +)	-4.605	0.149	0.000	0.503 (0.369, 0.680)***
Work (part-time)	-0.879	0.082	0.380	0.931 (0.790, 1.098)
Work (retired)	1.715	0.124	0.086	1.237 (0.973, 1.591)
Work (home duties)	0.435	0.076	0.663	1.034 (0.888, 1.204)
Work (student)	-2.853	0.138	0.004	0.674 (0.516, 0.891)**
Work (unemployed)	0.245	0.132	0.806	1.033 (0.795, 1.357)
Work (unable to work)	3.267	0.113	0.001	1.446 (1.160, 1.822)**
Edu (left school at 14)	-2.455	0.277	0.014	0.506 (0.280, 0.877)*
Edu (left school after 16)	-1.561	0.266	0.119	0.661 (0.372, 1.117)
Edu (left after 16 and studying)	-1.712	0.299	0.087	0.600 (0.320, 1.085)
Edu (trade/apprenticeship)	-2.077	0.274	0.038	0.566 (0.315, 0.974)*
Edu (certificate/diploma)	-1.445	0.261	0.148	0.685 (0.388, 1.152)
Edu (bachelor degree or higher)	-0.498	0.261	0.619	0.878 (0.497, 1.474)
Income ($20,000–$39,999)	-1.014	0.120	0.310	0.886 (0.693, 1.125)
Income ($40,000–$59,999)	0.864	0.120	0.387	1.110 (0.866, 1.413)
Income ($60,000–$79,999)	0.741	0.122	0.459	1.094 (0.851, 1.399)
Income ($80,000–$99,999)	1.118	0.125	0.264	1.151 (0.887, 1.484)
Income ($100,000–$119,999)	-0.618	0.123	0.536	0.927 (0.719, 1.188)
Income ($120,000–$139,999)	1.227	0.133	0.220	1.177 (0.898, 1.535)
Income ($140,000–$159,999)	0.971	0.141	0.331	1.146 (0.860, 1.524)
Income ($160,000 +)	1.151	0.119	0.250	1.147 (0.896, 1.457)
Family (young couple)	-0.966	0.126	0.334	0.885 (0.682, 1.148)
Family (young family)	-1.290	0.121	0.197	0.855 (0.662, 1.102)
Family (middle family)	-3.086	0.107	0.002	0.719 (0.573, 0.898)**
Family (mature family)	-2.585	0.103	0.010	0.766 (0.615, 0.950)**

	Estimate	Std.Error	p.value	IRR with CI
Family (mature couple or single)	-3.049	0.097	0.002	0.744 (0.604, 0.911)**
Family (other)	-1.742	0.152	0.082	0.767 (0.564, 1.050)
Gender (female)	-3.663	0.044	0.000	0.850 (0.777, 0.930)***
Barrier: Time (2)	-1.382	0.090	0.167	0.882 (0.736, 1.059)
Barrier: Time (3)	-5.354	0.079	0.000	0.656 (0.560, 0.768)***
Barrier: Time (4)	-7.965	0.075	0.000	0.551 (0.474, 0.639)***
Barrier: Time (5)	-11.369	0.074	0.000	0.431 (0.370, 0.501)***
Barrier: Safety (2)	0.061	0.059	0.951	1.004 (0.894, 1.128)
Barrier: Safety (3)	-0.306	0.080	0.760	0.976 (0.834, 1.146)
Barrier: Safety (4)	1.052	0.108	0.293	1.120 (0.906, 1.394)
Barrier: Safety (5)	1.166	0.139	0.244	1.177 (0.882, 1.592)
Barrier: Transportations (2)	-0.961	0.061	0.337	0.943 (0.838, 1.062)
Barrier: Transportations (3)	-4.300	0.071	0.000	0.737 (0.640, 0.850)***
Barrier: Transportations (4)	-2.042	0.082	0.041	0.845 (0.718, 0.997)*
Barrier: Transportations (5)	-2.220	0.094	0.026	0.812 (0.668, 0.992)*
Barrier: Entry cost (2)	-0.698	0.060	0.485	0.959 (0.852, 1.079)
Barrier: Entry cost (3)	-0.790	0.065	0.429	0.950 (0.834, 1.084)
Barrier: Entry cost (4)	-1.000	0.077	0.317	0.926 (0.793, 1.083)
Barrier: Entry cost (5)	-0.537	0.085	0.592	0.955 (0.804, 1.140)
Barrier: Facilities (2)	-0.618	0.067	0.537	0.959 (0.837, 1.100)
Barrier: Facilities (3)	-3.363	0.073	0.001	0.782 (0.672, 0.911)***
Barrier: Facilities (4)	-1.422	0.086	0.155	0.884 (0.739, 1.059)
Barrier: Facilities (5)	0.115	0.118	0.909	1.014 (0.793, 1.301)
Barrier: Not thinking (2)	-1.270	0.065	0.204	0.921 (0.807, 1.050)
Barrier: Not thinking (3)	-2.085	0.074	0.037	0.857 (0.735, 0.998)*
Barrier: Not thinking (4)	-2.520	0.088	0.012	0.800 (0.670, 0.958)*
Barrier: Not thinking (5)	-2.021	0.130	0.043	0.769 (0.592, 1.005)*
Year 2022	2.600	0.058	0.009	1.162 (1.035, 1.305)**
Year 2023	0.437	0.051	0.662	1.022 (0.923, 1.133)
NRM region 2	1.704	0.078	0.088	1.142 (0.978, 1.339)
NRM region 3	-0.973	0.180	0.331	0.839 (0.597, 1.213)
NRM region 4	-2.984	0.096	0.003	0.750 (0.620, 0.912)**
NRM region 5	-0.137	0.096	0.891	0.987 (0.818, 1.200)
NRM region 6	-2.036	0.097	0.042	0.821 (0.675, 1.006)*
NRM region 7	-1.383	0.209	0.167	0.749 (0.504, 1.153)
NRM region 8	-2.759	0.198	0.006	0.579 (0.395, 0.872)**
Observations				2,819

	Estimate	Std.Error	p.value	IRR with CI
Log likelihood				-8709.812
theta				0.973*** (0.027)
AIC				17,557.620
*Note: *p <0.05; **p <0.01; ***p <0.001*				

Table 4.
Negative binomial regression for visit count.

The model results highlight several significant demographic factors affecting visitation, including age, gender, employment, and household structure. For example, young singles visited the highest number of parks, while visitation decreased from the 35–44 age group onwards. Females also reported fewer visits than males. Among employment statuses, those unable to work recorded the highest number of visits. While this could be influenced by a smaller sample size for this group, it is an intriguing finding that prompts consideration of their motivations, such as seeking the therapeutic benefits of nature.

Interestingly, income had no significant effect on park visitation in our model, a finding that contrasts with some other studies [3, 8]. This may suggest the encouraging possibility that, in South Australia, access to national parks is not constrained by income level.

Regarding the barriers limiting park visitation, time was a major concern, which is consistent with previous research [9, 10]. Respondents who strongly agreed that time was a barrier reported only half the number of visits compared to those who strongly disagreed.

Interestingly, safety concerns made no difference in visitation frequency, while transportation difficulties had a slight negative effect. This latter point connects with our earlier findings for NRM 1, suggesting that people avoided the least-visited parks there not due to safety fears, but perhaps because of practical issues like poorer views, less-maintained trails, or longer travel distances. This paints a picture of visitors seeking recreational opportunities in beautiful, well-managed parks with good facilities that are not too far from home (e.g., within 100 km).

Another barrier, 'not knowing what to do in a national park', reduced park visitation by 8% to 23%. This is quite understandable and may result from limited information, uncertainty about available activities, or even poor park branding and promotion [11, 12].

6. A better life with parks around

Australia is one of only 17 megadiverse countries, hosting an extraordinary 10% of global biodiversity. This natural heritage offers significant benefits for human health and well-being, particularly when accessible from urban centres [13]. Green environments provide both avenues and motivation for physical activity, and research in Amsterdam demonstrated that each additional park within 1 km from home resulted in approximately 1.8 minutes more physical activity per week, contributing to a lower mortality risk [14]. The psychological benefits of nature exposure are also well-documented. In the UK, for example, increased tree cover has been associated with significant increases in perceived restorative benefit [15]. Similarly, a large Australian study of 46,786 adults found that exposure to 30% or more tree canopy, compared with 0–9% tree canopy, was associated with 31% lower odds of incident psychological distress [16].

Social connection is another key benefit. Natural environments in Australia provide valuable spaces for people to meet, play, and build relationships. A walk in the park with friends or family can improve mood, reduce stress, loneliness, and isolation for everyone involved, and improve overall health [17, 18].

This chapter's analysis of visitation patterns, land cover, and biodiversity across South Australia's NRM regions provides a local context for these benefits. While our study didn't directly link visitation with specific health outcomes, recent research has begun to quantify this value. By comparing health outcomes, one study estimated that access to biodiverse green spaces adjacent to Adelaide could generate AU $140 million in health-related savings for South Australia [19]. Scaled up across the country, this could reduce the nation's health bill by more than AU$2.1 billion annually [20].

This chapter suggests that providing a more diverse choice of biodiverse national parks near cities gives residents greater access to well-managed natural environments, where they can relax, recreate, and exercise. Combined with the evidence that visiting parks improves health, this presents a strong argument for maintaining and enhancing biodiversity in urban-proximate protected areas. Our analysis indicates that people seek well-maintained, beautiful, and accessible parks with native vegetation within a reasonable travel distance. In this regard, Adelaide's success as a National Park City offers a valuable model for other regions and cities worldwide.

Policy initiatives in South Australia are already aligning with these principles. In 2021, the state initiated the *Healthy Parks Healthy People SA* framework, with a vision to ensure all residents can experience the health benefits of connecting with nature [21]. Its action plans focus on promoting physical activity, mental health benefits, and the cultural value of the Country. The implementation of this policy coincided with a 16% increase in park visits in 2022, as shown in our data (**Table 4**). Although this rate did not increase further in 2023, the initial boost is promising. Such efforts, supported by the research cited here, not only benefit the environment but also contribute to a healthier society and a more resilient healthcare system.

7. Limitations

Our study has several limitations. First, our comparative analysis of biodiversity was confined to the most and least-visited parks within the Adelaide metropolitan area. This focused scope means the findings may not be generalizable to the differing ecological and social contexts of peri-urban NRMs. Furthermore, biodiversity is a multi-faceted concept that cannot be fully captured by a single metric. While the Shannon index provides a useful snapshot, future research could build on our work by employing a more comprehensive suite of ecological indicators to explore the nuanced relationship between specific ecosystem characteristics and park visitation across all regions.

Second, the survey data did not capture the frequency of visits to each specific park, which prevented us from analysing visitor preferences in more detail. We were only able to assess the overall concentration of visits.

Finally, the relationship between park visitation and human well-being is complex. While our study and others show a positive link, it is difficult to draw a direct causal conclusion that living near more national parks makes people healthier. Nonetheless, the connection is clear, and the public's appreciation for these spaces is evident.

8. Conclusion

This chapter highlights the important role of biodiverse natural areas in and around cities. Our analysis of park visitation in South Australia reveals a key distinction between urban and peri-urban residents, showing that residents in the Adelaide metropolitan area explore a more diverse range of parks. While our investigation into the drivers of this pattern pointed to a complex interplay of factors, it was clear that visitor choices are influenced by preferences for specific native vegetation types, well-maintained facilities, and accessibility. The decision to visit a park is also shaped by sociodemographic factors and practical barriers like time constraints. These findings underscore the value of investing in a diverse portfolio of high-quality, biodiverse, and well-managed natural spaces where people live. By doing so, we can encourage engagement with nature, which in turn fosters public health, social connection, and more resilient and vibrant communities.

Acknowledgments

We acknowledge Agamoni Majumder for data preparation. The data was provided by the South Australian Department for Environment and Water (DEW) as part of an ongoing collaboration.

Author details

Jianqiang Huang* and Patrick O'Connor
University of Adelaide, Adelaide, Australia

*Address all correspondence to: keon.huang@adelaide.edu.au

References

[1] Fromm O. Ecological structure and functions of biodiversity as elements of its total economic value. Environmental and Resource Economics. 2000;**16**(3):303–328

[2] Weber J, Sultana S. Why do so few minority people visit national parks? Visitation and the Accessibility of "America's best idea. Annals of the Association of American Geographers. 2013;**103**(3):437–464

[3] Balmford A, Green JMH, Anderson M, Beresford J, Huang C, Naidoo R, et al. Walk on the Wild Side: Estimating the Global Magnitude of visits to protected areas. PLOS Biology. 2015;**13**(2):e1002074

[4] Crespo-Cebada E, Díaz-Caro C, Robina-Ramírez R, Sánchez-Hernández MI. Is biodiversity a relevant attribute for assessing natural parks? Evidence from Cornalvo Natural Park in Spain. Forests. 2020;**11**(4):410

[5] Eyster HN, Naidoo R, Chan KMA. Not just the big five: African ecotourists prefer parks brimming with bird diversity. Animal Conservation. 2023;**26**(4):428–442

[6] Tanner S, Lupi F, Garnache C. Estimating visitor preferences for recreation sites in wildfire prone areas. International Journal of Wildland Fire. 2022;**31**(9):871–885

[7] Lis A, Pardela Ł, Iwankowski P. Impact of vegetation on perceived safety and preference in City Parks. Sustainability. 2019;**11**(22):6324

[8] Xiao X, Lee KJ, Larson LR. Who visits U.S. national parks (and who doesn't)? A national study of perceived constraints and vacation preferences across diverse populations. Journal of Leisure Research. 2022;**53**(3):404–425

[9] Crompton JL, Kim SS. Temporal changes in perceived constraints to Visiting State Parks. Journal of Leisure Research. 2004;**36**(2):160–182

[10] Gong Y, Mamat Z, Shi L, Liu F. Restorative effects of park visiting on physiology, psychology, and society and the factors influencing park visiting. Sustainability. 2023;**15**(1):841

[11] Ramsay G, Dodds R, Furtado D, Mykhayletska Y, Kirichenko A, Majedian M. The barriers to millennials visiting Rouge Urban National Park. Sustainability. 2017;**9**(6):904

[12] Franceschinis C, Swait J, Vij A, Thiene M. Determinants of recreational activities choice in protected areas. Sustainability. 2022;**14**(1):412

[13] Dielenberg J, Bekessy S, Cumming GS, Dean AJ, Fitzsimons JA, Garnett S, et al. Australia's biodiversity crisis and the need for the Biodiversity Council. Ecological Management & Restoration. 2023;**24**(2–3):69–74

[14] Remme RP, Frumkin H, Guerry AD, King AC, Mandle L, Sarabu C, et al. An ecosystem service perspective on urban nature, physical activity, and health. Proceedings of the National Academy of Sciences. 2021;**118**(22):e2018472118

[15] Hassall C, Nisbet M, Norcliffe E, Wang H. The potential health benefits of urban tree planting suggested through immersive environments. Land. 2024;**13**(3):290

[16] Astell-Burt T, Feng X. Association of urban green space with mental health and general health among adults in Australia. JAMA Network Open. 2019;**2**(7):e198209

[17] Anderson AR. Enjoying nature, exercise, social interaction, and affect: A daily diary study. Journal of Health Psychology. 2022;**27**(4):890–900

[18] Robinson JM, Breed AC, Camargo A, Redvers N, Breed MF. Biodiversity and human health: A scoping review and examples of underrepresented linkages. Environmental Research. 2024;**246**:118115

[19] Loch A, Sexton S, Maclean J, O'Connor P, Adamson D, Scholz G. Increased monetary equity and health wellbeing benefits for marginal urban socioeconomic groups from access to green space. Urban Forestry & Urban Greening. 2024;**102**:128576

[20] O'Connor P, Loch AJ, Maclean J. The Conversation. 2024. 'A Dose of Nature': Each Time You Visit a National Park, You Save the Health Budget Almost $100. Available from: https://theconversation.com/a-dose-of-nature-each-time-you-visit-a-national-park-you-save-the-health-budget-almost-100-243916 [Accessed: 2025-February-18]

[21] Department for Environment and Water. Healthy Parks Healthy People SA 2021 - 2026. 2021. Available from: https://www.environment.sa.gov.au/topics/park-management/statewide-park-strategies/healthy-parks-healthy-people [Accessed: 2025-September-27]

Chapter 3

Rethinking Land Use in Watershed Regions: Istanbul's Lessons for Sustainable Futures

Ozge Celik Yilmaz

Abstract

The research explores the potential and limitations of ecosystem-based land use planning in rapidly urbanizing contexts, using Istanbul's ecologically sensitive waterscapes – including the Riva River Basin, Ömerli Reservoir, and Şile Region – as case studies. Through a critical examination of land use changes over the last two decades, supported by remote sensing data and spatial policy analysis, the research reveals the growing mismatch between ecological needs and current planning practices. The findings highlight how urban expansion, institutional fragmentation, and insufficient environmental regulation have led to the degradation of vital wetland ecosystems. Despite the existence of strategic planning documents such as Istanbul Vision 2050, implementation remains weak in protecting natural systems. The research argues for a paradigm shift toward ecosystem-based, adaptive, and participatory planning approaches. It concludes by offering policy recommendations for integrating nature-based solutions, green infrastructure, and watershed-scale governance into formal land use planning frameworks, contributing to more sustainable and climate-resilient urban futures.

Keywords: ecosystem-based planning, urban land use change, watershed management, nature-based solutions, Istanbul urbanization

1. Introduction

In the face of escalating climate crises, biodiversity loss, and rapid urban expansion, land-use planning is undergoing a profound transformation worldwide. Increasingly, planners and policymakers are turning toward ecosystem-based approaches and nature-based solutions (NBS) to reconcile the demands of urban growth with the imperative of ecological resilience [1–3]. These approaches prioritize the integrity of watersheds, the protection of biodiversity, and the restoration of ecosystem functions as central pillars of spatial governance. Globally, there is growing recognition that conventional, sectoral planning models – often rigid and reactive – are insufficient to address the complex socio-ecological challenges of the twenty-first century [4, 5]. The urgency of this shift is particularly evident in urban regions, where freshwater and wetland ecosystems are under acute pressure from land conversion and environmental degradation [6–8].

Watershed planning has emerged as a critical framework within this paradigm shift, integrating hydrological, ecological, and socio-economic dimensions to manage land and water resources in a holistic manner [9–12]. In the context of the climate crisis, watershed-based approaches are essential for maintaining water security, regulating urban microclimates, preventing floods, and safeguarding biodiversity [9–12]. Changing precipitation regimes, extreme weather events, and rising sea levels intensify the vulnerability of watershed systems, making adaptive, cross-scale governance an urgent necessity [9–15]. Yet, despite their proven potential, the adoption of ecosystem-based and watershed-focused planning strategies remains uneven, particularly in rapidly urbanizing regions of the Global South.

In Türkiye, and especially in Istanbul, these challenges are magnified. The city's northern waterscapes – including the Riva River Basin, Ömerli Reservoir, and the rural–coastal fringe of Şile – function as vital ecological corridors and primary water catchments, supplying millions of residents with drinking water while providing climate regulation, flood mitigation, and biodiversity conservation services [16, 17]. Over the past two decades, however, these landscapes have been subjected to intensive land-use change driven by urban sprawl, infrastructure expansion, and fragmented governance. National and local planning instruments – such as the Environmental Law, watershed protection regulations, the Istanbul Environmental Master Plan (2009), and the Istanbul Vision 2050 Strategy (2021) – formally recognize the need to protect these areas [16, 17]. Yet, their implementation is hampered by overlapping mandates, weak enforcement, and political contestation between central and municipal authorities.

This chapter argues that Türkiye's current planning system is ill-equipped to address the socio-ecological pressures confronting urban watershed regions. Institutional fragmentation, inconsistencies between regulatory frameworks, and the erosion of public trust – exacerbated by cases of protected land reclassification for development – have undermined governance integrity and environmental justice. Against this backdrop, the chapter focuses on the Riva–Ömerli–Şile corridor as a critical case for examining the interplay between ecological vulnerability, governance structures, and land use change.

The chapter builds its argument on a thematic analysis of institutional shortcomings, policy gaps, and spatial vulnerabilities – assessed through both land-use change mapping (2000–2024) and ecological risk evaluation – to reveal the growing mismatch between ecological thresholds and planning outcomes. By situating this case within broader debates on sustainable urban development and climate adaptation, the study underscores the need for a paradigmatic shift toward ecosystem-based, adaptive, and participatory planning that integrates NBS into formal spatial governance.

Against this backdrop, this chapter investigates the Riva–Ömerli–Şile corridor as a critical case for understanding the interplay between ecological vulnerability, governance structures, and land-use change. The analysis addresses three research questions:

How have land-use changes in Istanbul's water-sensitive landscapes evolved over the past two decades, and what ecological impacts have emerged?

To what extent do existing planning instruments and strategies address these transformations?

What planning and policy mechanisms can support the integration of ecosystem-based and NBS into formal land-use planning frameworks?

By addressing these questions, the chapter contributes to academic and policy debates on ecological resilience, green infrastructure, and watershed-scale governance. The findings aim to inform actionable strategies for integrating ecosystem-based planning, NBS, and adaptive governance into urban and regional planning processes – repositioning Istanbul's waterscapes as pillars of a climate-resilient urban future.

2. Ecosystem-based land use planning: Conceptual workground

Ecosystem-based land use planning (EBLUP) has emerged as a critical framework to overcome the shortcomings of conventional spatial planning paradigms. Traditional approaches have often prioritized zoning efficiency, economic growth, and sectoral infrastructure over ecological integrity, leading to the neglect of thresholds, watershed-scale interactions, and the systemic connectivity of natural systems [1–6, 18]. By contrast, EBLUP emphasizes the integration of ecological processes, hydrological dynamics, and resilience principles into spatial governance. Its central aim is to ensure that development remains within the carrying capacity of landscapes while safeguarding biodiversity, maintaining watershed integrity, and securing long-term ecosystem services [18].

On the other hand, watersheds represent a natural unit of ecological governance, making them a cornerstone of EBLUP. Unlike administrative boundaries, which are often politically defined and fragmented, watersheds encompass the hydrological, ecological, and socio-economic processes that directly affect land and water systems [6–15]. Poorly managed upstream activities – such as deforestation, agricultural intensification, and urban sprawl – trigger cascading downstream impacts, including flood risk, water quality deterioration, and habitat loss [12–15]. For example, watershed degradation has been shown to reduce groundwater recharge, accelerate soil erosion, and increase nutrient runoff, directly affecting both human settlements and ecological resilience [7].

Moreover, global assessments by the IPCC Sixth Assessment Report [19] confirm that watershed ecosystems – particularly wetlands, riparian buffers, and floodplains – are critical for climate adaptation, disaster risk reduction, and achieving Sustainable Development Goals (SDGs). Specifically, EBLUP directly contributes to SDG 6 (Clean Water and Sanitation), SDG 13 (Climate Action), and SDG 15 (Life on Land) by maintaining freshwater quality, regulating climate-sensitive ecosystems, and conserving terrestrial habitats.

Furthermore, international policy frameworks reinforce the role of EBLUP. The UN SDGs (2015), the Convention on Biological Diversity (CBD), and the EU Green Infrastructure Strategy emphasize the protection and restoration of ecological networks as essential for building urban resilience. These frameworks advocate measures such as riparian buffer protection, floodplain restoration, and wetland conservation, aligning spatial planning with ecological thresholds [19]. However, practical applications demonstrate how these principles can be embedded in urban governance. Cities such as Copenhagen, Singapore, and Amsterdam have operationalized EBLUP through statutory plans that integrate green–blue infrastructure, ecological zoning, and floodplain restoration, supported by GIS-based vulnerability assessments and participatory decision-making processes [8–15]. These cases illustrate the potential of coupling ecological science with planning policy to build adaptive and climate-resilient cities.

On the other hand, GIS-based vulnerability assessments and ecosystem service trade-off analyses also guide decision-making. For instance, integrated land–water models have been used to locate high-risk zones for nonpoint source pollution and assess land use transitions that balance agricultural productivity with water quality protection [20]. Spatial statistical methods – such as kriging, semivariance analysis, and machine learning classifiers (XGBoost, Random Forest) – further improve the accuracy of land use/land cover (LULC) prediction and vulnerability mapping [21–29]. These approaches allow more precise delineation of priority conservation areas, particularly in rapidly urbanizing contexts where ecological thresholds are easily exceeded [21–29]. However, a critical dimension of EBLUP is its integration with adaptive governance and multiscalar coordination. Research shows that misalignments between local land use policies and basin-level hydrological realities often result in fragmented management and ecological degradation [30, 31]. Effective implementation requires cross-sectoral collaboration, horizontal alignment across municipalities, and vertical coordination between central and local governments. For instance, integrating watershed-scale monitoring, restoring ecological corridors, and embedding climate-resilient zoning regulations can significantly reduce nonpoint source pollution, maintain habitat connectivity, and enhance resilience to climate extremes [32–34]. Multilevel governance frameworks also enable participatory mechanisms, such as community-based watershed councils, which ensure that local knowledge complements scientific data in land management.

Moreover, EBLUP is particularly urgent in rapidly urbanizing watershed regions of the Global South, where competing land-use pressures threaten both biodiversity and water security. Evidence from South Africa, India, and Mediterranean regions demonstrates that interventions such as riparian reforestation, soil conservation practices, and wetland restoration significantly improve infiltration, reduce erosion, and enhance water quality [7, 8, 11, 13, 15, 18, 21–26, 28, 29, 31, 34–36]. Yet, the success of such interventions is highly contingent on governance structures: while multilevel governance can enable coordination, policy fragmentation often leads to conflicting land-use priorities [37].

Furthermore, scenario-based modeling supports the design of adaptive strategies. Coupling LULC change models with hydrological and climate models enables projections of future risks and the identification of intervention zones [22–26]. Importantly, recent studies incorporate socio-economic drivers – such as population growth, migration, and land tenure regimes – to better reflect the complexity of real-world land-use decisions [23].

However, the literature converges on four core principles for effective ecosystem-based land-use planning:

- Data-driven: Utilize high-resolution RS, GIS, and modeling tools for continuous ecological monitoring.

- Governance-aware: Align policies at local, regional, and national levels through multilevel coordination.

- Service-oriented: Explicitly account for ecosystem service trade-offs in land allocation decisions.

- Participatory: Engage communities, NGOs, and stakeholders in codesigning strategies and monitoring implementation.

Together, these principles provide a robust framework for balancing development pressures with ecological limits.

Taken together, EBLUP bridges ecological science, hydrological modeling, and spatial governance, offering a holistic strategy to mitigate ecological risks and enhance socio-economic resilience. Its application is particularly urgent in watershed regions where urbanization pressures intersect with climate-induced vulnerabilities. By combining scientific tools, governance frameworks, and participatory mechanisms, EBLUP offers a scalable approach for sustainable territorial development. While global cases demonstrate its potential, the Turkish context – characterized by centralization, policy fluidity, and contested ecological zones – remains underexplored. This chapter thus extends the debate by examining how EBLUP can be adapted and operationalized in Istanbul's northern watersheds, where ecological thresholds and governance gaps converge. While ecosystem-based approaches have been widely discussed at the global scale, their translation into national and local planning frameworks remains uneven. In Turkey, this gap is particularly visible in watershed regions such as Riva, Ömerli, and Şile.

3. Fragmented planning and governance challenges in Türkiye

In the Turkish context – particularly in ecologically sensitive regions such as the Riva River Basin, Ömerli Reservoir catchment, and Şile coastal landscapes – EBLUP offers a pathway to reconcile urban growth with ecosystem protection. These areas face mounting pressures from speculative housing, tourism development, and infrastructure expansion, all within critical drinking water protection zones. By integrating ecosystem service valuation, hydrological modeling, and participatory governance, EBLUP can inform more adaptive, resilience-oriented spatial plans that address both SDGs and local livelihood needs.

In Türkiye, and especially in Istanbul, the relevance of EBLUP and watershed-based management is particularly pronounced. The city's northern waterscapes – including the Riva River Basin, Ömerli Reservoir, and the rural–coastal fringe of Şile – function as primary water catchments and ecological corridors, supplying millions of residents with drinking water while providing flood mitigation, climate regulation, and biodiversity conservation services [16, 38]. However, over the past two decades, these landscapes have undergone significant land-use change due to urban sprawl, infrastructure expansion, and fragmented governance [37]. Despite formal recognition in national and municipal plans – such as the Environmental Law, Istanbul Environmental Master Plan (2009), and Istanbul Vision 2050 Strategy (2021) – implementation has been hindered by overlapping mandates, weak enforcement, and political contestation between central and local authorities [16, 17, 37]. Integrating EBLUP with robust watershed governance in these regions is thus critical to safeguard ecological thresholds, ensure long-term water security, and strengthen climate resilience.

Although Türkiye's spatial planning system appears to be supported by a comprehensive legal and institutional framework, its capacity to manage ecologically sensitive areas – particularly watershed zones – remains structurally weak. The coexistence of overlapping legislative instruments, a governance model that is nominally decentralized but functionally hierarchical, and a mismatch between planning scales has produced a fragmented and often contradictory approach to land use management. This fragmentation becomes most visible in metropolitan contexts such as Istanbul, where ecological boundaries – river basins, forest zones, wetlands – rarely align with administrative borders or statutory planning jurisdictions [4].

However, a central source of fragmentation lies in the proliferation of plans and planning authorities operating across multiple spatial scales. National strategic documents, regional development plans, metropolitan environmental plans, and district-level zoning plans are often prepared in parallel without effective horizontal or vertical coordination. In watershed regions, land use decisions can simultaneously fall under the jurisdiction of the Ministry of Environment, Urbanization and Climate Change; the Istanbul Metropolitan Municipality; district municipalities; and specialized water or forestry agencies, such as the Istanbul Water and Sewerage Administration and the General Directorate of Forestry [39]. This institutional layering fosters planning incoherence and policy contradictions, particularly when political tensions between central and local administrations are high.

Moreover, in Istanbul, this misalignment often results in plans that compete rather than complement each other. For example, while the Istanbul Environmental Master Plan (2009) designates key basins such as Riva, Ömerli, and Şile as "ecologically prioritized," district-level zoning revisions have opened parts of these same basins to residential or tourism development [16, 17]. Similarly, the Ministry's large-scale infrastructure projects – such as new highways or energy corridors – have overridden municipal conservation plans, illustrating the vulnerability of ecological priorities in the face of national development agendas [16, 17]. Over the past decade, Türkiye has witnessed a gradual erosion of environmental protections through legislative amendments and regulatory exceptions. The reclassification of Special Environmental Protection Areas (SEPAs), the use of presidential decrees to grant exceptional development rights, and frequent revisions to construction and environmental laws have collectively weakened the enforceability of conservation principles. The shift to the presidential system in 2018 has further centralized decision-making, enabling top-down interventions that override locally prepared plans. This legal fluidity creates an unstable regulatory environment, where rules are subject to rapid change and enforcement is inconsistent. In practice, statutory buffer zones around rivers and reservoirs – defined in the Water Pollution Control Regulation (2004) and the Regulation on Protection of Drinking Water Basins (1998/2004) – are routinely compromised by rezoning decisions, infrastructure alignments, and illegal construction. Field surveys in the Riva and Ömerli basins reveal numerous cases where housing developments, tourism facilities, and road projects have encroached into absolute or short-distance protection zones, directly contravening national regulations. The Spatial Plans Construction Regulation (2014) formally requires that ecological thresholds be considered in planning, yet the provision remains nonbinding ("should be considered") and lacks operational guidance. As a result, ecological data is rarely integrated into zoning maps or development control processes in a systematic way. Strategic documents such as Istanbul Vision 2050 and the National

Policy/regulation	Scope	Application area	Critical assessment
Water Pollution Control Regulation (2004)	Defines 10 m (absolute) and 100 m (short-distance) buffer zones	Stream and tributaries	Weak integration into spatial plans; often overridden by zoning decisions
Regulation on the Protection of Drinking Water Basins (1998/2004)	Four-zone system for reservoirs and catchments	Basin	Strong in theory; weak in enforcement; illegal
Spatial Plans Construction Regulation (2014)	Requires consideration of ecological thresholds	Nationwide urban planning processes	Nonbinding language ("should be considered"); lacks data and guidance
Istanbul Environmental Master Plan (2009) / Vision 2050	Marks Riva, Şile, Ömerli as "ecologically prioritized"	Istanbul metropolitan macroform	Strong ecological rhetoric, but implementation depends on central approvals
UBSEP – National Biodiversity Strategy (2018–2028)	Promotes biodiversity and ecosystem services	Nationwide	Low integration with planning; weak binding authority
DSİ Stream Rehabilitation Projects	Flood prevention via engineering works	Rural streams across Istanbul	Hard infrastructure undermines eco-hydrological integrity; NBS underused

Table 1.
Policy and regulatory framework for watershed and riparian planning in Türkiye.

Biodiversity Strategy (2018–2028) promote ecosystem-based approaches and NBS, but these remain largely aspirational in the absence of binding implementation mechanisms, performance indicators, and dedicated funding streams [16, 40] (**Table 1**).

Moreover, planning outcomes are often shaped by the dominance of sectoral, engineering-led approaches. The State Hydraulic Works, for instance, continues to prioritize hard-infrastructure flood control measures – such as concrete channelization – over eco-hydrological restoration in river corridors. This sectoral bias not only undermines the ecological integrity of watersheds but also perpetuates a reactive, project-based model of intervention rather than a preventive, landscape-scale planning approach [41].

Despite the presence of these regulatory tools, implementation remains inconsistent and fragmented. Field evidence from the Riva and Ömerli basins reveals widespread violations of buffer zone regulations, highlighting the absence of effective enforcement mechanisms. Strategic documents, while emphasizing ecosystem-based approaches and NBS, often fail to translate into concrete, enforceable spatial planning measures. Moreover, sectoral dominance – particularly engineering-led flood control by the State Hydraulic Works – undermines integrated watershed management [16, 17, 40, 41].

Furthermore, enforcement agencies often lack the resources, authority, or political backing to prevent encroachment in high-sensitivity zones. The coexistence of ambitious strategic visions and short-term, project-driven planning produces a chronic planning–practice gap. In watershed regions like Riva and Ömerli, this gap manifests as uncoordinated urban growth, habitat fragmentation, and declining water quality – trends that directly undermine ecological resilience and increase vulnerability to climate-induced hazards.

Addressing these governance failures requires more than technical planning updates; it demands institutional reform to ensure legal coherence, strengthen multilevel governance, and restore public trust in environmental decision-making. Without such systemic change, Istanbul's watershed regions will remain locked in a cycle of ecological degradation and reactive policy responses.

3.1 Landscape risks in Istanbul's watershed regions: The case of Riva, Ömerli, and Şile regions

Istanbul's rapid urbanization over the past two decades has significantly transformed its ecologically sensitive northern periphery, where key watershed regions such as the Riva River Basin, Ömerli Reservoir, and the Şile coastal fringe are located. These landscapes are critical to the city's ecological balance, water security, and climate regulation, functioning as primary freshwater sources, biodiversity corridors, and buffers against climate-induced hazards. Yet, they have become focal points for speculative urban expansion, fragmented infrastructure development, and policy inconsistency (**Table 2**).

Although national planning documents, such as the Integrated Basin Management Strategy and Climate Change Action Plans, reference ecological planning principles, implementation on the ground is limited and inconsistent [16, 17, 40, 41]. The fragmented and sectoral structure of planning institutions hampers the integration of ecological data into land-use decisions, particularly in dynamic metropolitan areas like Istanbul. For instance, river basins such as Riva and Ömerli, despite being recognized as ecological assets and water supply sources, continue to face pressures from uncoordinated urban expansion and infrastructure development.

Furthermore, the lack of institutionalized mechanisms to link ecological monitoring with spatial planning tools – such as land suitability analysis, ecosystem service mapping, and flood risk modeling – limits the ability of planners to anticipate

Policy/regulation	Intended goal	Implementation status	Gap/conflict
Water Pollution Control Regulation (2004)	Define 10–100 m riparian protection zones	Illegal construction, road projects along Riva	Enforcement deficit – weak monitoring
Regulation on the Protection of Drinking Water Basins (1998/2004)	Prohibit construction in absolute/short zones	Zoning plan inconsistencies	Law–plan conflict, institutional weakness
Spatial Plans Construction Regulation (2014)	Incorporate ecological thresholds into planning	Ecological data often ignored	Guidance/data deficit
Istanbul Vision 2050	Promote NBS and ecosystem-based planning	No binding strategies or projects	Strategy–implementation gap
DSİ Stream Rehabilitation Projects	Control flooding	Hard engineering dominance in Riva and Şile	Ecological–sectoral mismatch

Table 2.
Policy gap analysis: Şile, Riva, Ömerli.

and manage environmental risks. While some pilot initiatives (e.g., TÜBİTAK-supported ecological corridors, urban green networks) have been attempted, these remain isolated and insufficiently scaled [38].

3.1.1 Riva stream corridor

Historically dominated by mixed deciduous forests, extensive wetlands, and meandering riparian corridors, the Riva Basin has been subject to incremental but persistent land-use conversion. Located within a designated drinking water protection zone, the basin has experienced both authorized and unauthorized construction, facilitated by zoning plan amendments and overlapping jurisdictional mandates.

- Key drivers of risk include unplanned tourism facilities, second-home developments, and road construction projects that bisect ecological corridors.

- Ecological consequences include riparian vegetation loss, habitat fragmentation, and increased impervious surface coverage, which, in turn, elevate peak flood discharges and sediment loads in downstream settlements [39].

- Field surveys reveal that short-distance protection zones – intended to be development-free – are increasingly encroached upon, indicating regulatory enforcement gaps and a weak culture of compliance.

To address these pressures, basin authorities could prioritize riparian buffer restoration and introduce permit systems limiting tourism facilities within short-distance protection zones. Developing a dedicated Riva Basin Management Unit, supported by regular ecological audits and transparent reporting, would enhance compliance and prevent further corridor fragmentation.

3.1.2 Ömerli basin

As Istanbul's largest surface water source, the Ömerli catchment is governed by a four-zone protection hierarchy under national regulation. While the absolute protection zone remains largely intact, medium and long-distance zones have seen significant low-density residential expansion since the early 2000s.

- Pressure points include mining activities, agricultural intensification, and transportation corridors (notably the Northern Marmara Motorway), which increase sediment inflows and nutrient loading into the reservoir.

- Water quality monitoring [41] indicates rising concentrations of nitrogen and phosphorus, attributed to both point sources (e.g., septic leakage) and diffuse sources (e.g., fertilizer runoff).

- Climate projections suggest that prolonged summer droughts, combined with reduced forest cover, will exacerbate eutrophication risks and diminish the catchment's resilience to extreme events [38].

Policy responses should focus on zero-tolerance enforcement of the four-zone hierarchy, combined with incentive schemes for farmers to adopt erosion-reducing practices. Integrating hydrological modeling into zoning revisions would allow planners to anticipate nutrient loading and design adaptive runoff controls.

3.1.3 Şile coastal fringe

The Şile district, straddling forested uplands and the Black Sea coast, functions as a dual ecological zone – serving both as a biodiversity hotspot and a cultural landscape supporting traditional fishing and agriculture. In recent years, the area has become a focal point for mass tourism development, second-home proliferation, and coastal infrastructure expansion.

- Shoreline hardening (e.g., seawalls, piers) has disrupted natural sediment dynamics, increasing erosion rates in certain stretches while causing accretion in others.

- The inland forest–coast transition zone is increasingly fragmented by tourism, roads, and gated communities, reducing wildlife mobility and altering hydrological connectivity.

- Seasonal population surges (up to a 300% increase in summer months) place additional stress on water supply systems, waste management infrastructure, and coastal ecosystems.

Introducing coastal setback lines, regulating seasonal visitor capacity, and promoting eco-tourism standards could help reconcile recreational demand with shoreline stability and habitat continuity.

3.1.4 Shared challenges and governance gaps

Across all three basins, the same systemic issues emerge: Institutional fragmentation, regulatory loopholes, and insufficient civic oversight. While strategic frameworks such as the Istanbul Environmental Plan (2009) and Vision 2050 emphasize ecological priorities, they lack binding performance indicators, enforcement mechanisms, and cross-sector coordination. In the absence of integrated, basin-scale planning and participatory governance, land-use changes in these regions will continue to undermine water security, biodiversity, and climate resilience [16, 17].

Moreover, the northern watershed regions of Istanbul exhibit a high-risk profile defined by high exposure to urbanization pressures, low adaptive capacity due to institutional fragmentation, and the erosion of ecosystem services. Without a transition toward basin-scale, participatory, and ecosystem-informed planning, these waterscapes risk becoming irreversible casualties of metropolitan growth. The case examples presented here underscore how governance failures translate into tangible ecological degradation, highlighting the urgency of reforming both regulatory frameworks and planning practices.

However, integrating EBLUP with watershed-based management in Istanbul's northern ecological corridors offers a direct pathway to strengthen climate resilience, safeguard water security, and prevent irreversible biodiversity loss. Policymakers must move beyond fragmented, sectoral approaches and establish multilevel governance frameworks that enable coordination between central, municipal, and basin-level authorities [37]. This requires: Stronger regulatory enforcement to ensure that environmental thresholds – such as buffer zones in drinking water catchments – are strictly protected; integration of NBS into statutory planning instruments, ensuring that wetland restoration, forest conservation, and riparian buffer creation are mandatory components of master plans; participatory watershed councils that include municipalities, regional agencies, NGOs, and community stakeholders, enabling co-produced knowledge and shared responsibility; GIS-based monitoring systems for real-time tracking of land use changes, hydrological dynamics, and ecosystem service provision. Such measures would not only align Istanbul's planning system with global sustainability frameworks like the SDGs and EU Green Infrastructure Strategy, but also address the city's pressing socio-ecological vulnerabilities by making watershed governance the central organizing principle for spatial planning.

Embedding basin report cards and open ecological dashboards into municipal and basin authority workflows would enhance accountability and public oversight, while independent environmental ombudsman mechanisms could strengthen the legal recourse for violations. Practical interventions include establishing basin-scale coordination platforms with clear mandates, embedding nature-based flood control into stream rehabilitation programs, and allocating performance-based budgets to agencies responsible for watershed enforcement.

4. Methodological approach

This methodological framework was designed to capture both the ecological and governance dimensions of watershed management in Northern Istanbul, specifically focusing on the Şile Stream Corridor, the Ömerli Basin, and the Riva River Basin. These areas, representing critical water supply and biodiversity corridors, also reflect the pressures of urbanization and fragmented governance structures.

This research employs a multiscalar, mixed-methods framework to investigate ecological vulnerabilities and urbanization pressures in environmentally sensitive areas of Istanbul, focusing on forest ecosystems, protected zones, river corridors, and drinking water basins on the Anatolian side (Şile Region, Ömerli Basin, and Riva River Basin). The primary objective is to assess how land-use dynamics and institutional governance structures contribute to ecological degradation and water quality deterioration under the ongoing climate crisis.

The methodology consists of three interrelated analytical stages:

4.1 Stage 1 – land use and land cover change (LULC) analysis (1990–2018)

Objective: Detect long-term spatial transformations across forest, agriculture, bushland, and settlement classes.

Data sources: Multitemporal satellite imagery (Landsat, Sentinel) and CORINE Land Cover datasets.

Processing:

- Preprocessing and classification were conducted in Google Earth Engine (GEE) using Random Forest (RF) and Support Vector Machine (SVM) algorithms.
- Accuracy assessment employed ground truth data, confusion matrices, and Kappa coefficients.

Change detection:

- Net gain/loss of forests, agriculture, wetlands, and built-up areas were calculated.
- Landscape metrics (fragmentation index, edge density) were computed using FRAGSTATS to detect habitat fragmentation trends.

LULC formulas applied:

Net change (ha)

Net change = A_{t2} – A_{t1}

Rate of change (%)

Rate of change = ((A_{t2} – A_{t1})/A_{t1}) × 100

Annual rate (%/year) (CAGR)

Annual rate = (A_{t2}/A_{t1})^(1/(t2 – t1)) – 1

Where:

A_{t1} = area (ha) at the beginning year

A_{t2} = area (ha) at the ending year

t2 – t1 = number of years between two periods

GIS integration: Buffer zones (0–1 km, 1–3 km, 3–5 km) were generated around rivers and reservoirs in ArcGIS Pro to quantify settlement densities (linked to **Table 8**).

4.2 Stage 2 – institutional governance networks mapping

Scope: National (Ministry of Environment, DSİ, İSKİ) and local (IMM, district municipalities) actors.

Policy review: Strategic documents, including the Istanbul Environmental Plan (2009), Vision 2050 Strategy, and the Istanbul Green City Action Plan, were analyzed to assess planning-policy alignment with ecological priorities.

Assessment criteria:

- Alignment between policy goals and observed land-use outcomes.
- Institutional overlaps, gaps, and conflicts.
- Degree of multilevel coordination in watershed protection.

Outcome: Governance mapping revealed institutional fragmentation and a lack of basin-scale enforcement, particularly in the Ömerli and Riva basins, reinforcing spatial vulnerabilities.

By cross-referencing the results with Vision 2050, the Green City Action Plan, and the Environmental Plan, the methodology ensures that findings are directly embedded within Istanbul's strategic planning framework. This allows the results (**Tables 8–9**) not only to present descriptive statistics but also to inform actionable governance reforms.

4.3 Limitations

Despite the robustness of this mixed-methods design, several limitations should be acknowledged.

- The spatial resolution of the satellite data (10–30 m) constrained the detection of fine-scale landscape changes, particularly in riparian zones and narrow corridors.

- Temporal coverage was limited by the availability of cloud-free images; some intermediate years could not be fully represented, potentially affecting the continuity of LULC trends.

- Governance data relied partly on policy documents and stakeholder interviews; access to internal reports and enforcement records was restricted, which may have led to partial representations of institutional dynamics.

4.4 Future research directions

Building on the present framework, future studies could:

- Combine participatory mapping and scenario workshops with local stakeholders to coproduce policy responses and evaluate trade-offs between conservation and development.

- Develop spatially explicit models to simulate the effects of alternative enforcement or restoration strategies on forest integrity and water quality.

- Extend the governance analysis to include financial and legal instruments (e.g., payment for ecosystem services, basin-scale regulatory tools) to better understand leverage points for reform.

5. Rethinking land use in watershed regions in Türkiye

The northern landscapes of Istanbul constitute one of the most strategically important ecological and hydrological systems in Türkiye, serving as a critical source of drinking water, biodiversity, and climate regulation for the metropolitan region. Encompassing major basins such as Elmalı, Ömerli, and Darlık, and vital stream corridors including Riva, Şile, and Ağva, these areas form a complex mosaic of

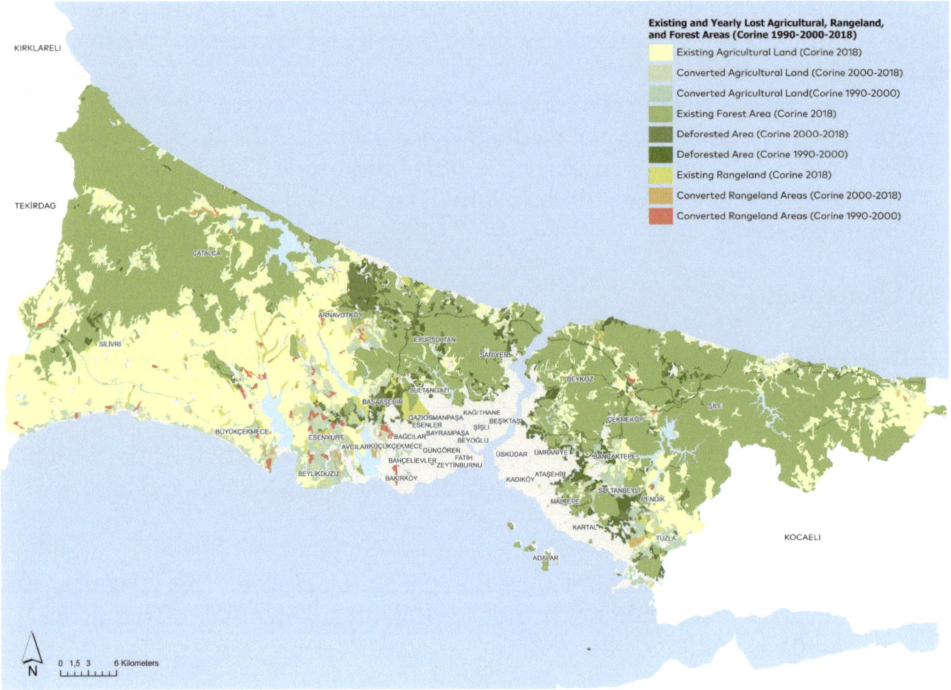

Figure 1.
Land use and land cover change (1990–2000–2018).

natural ecosystems, agricultural lands, and expanding settlements. Their ecological functions – ranging from water provisioning, flood regulation, carbon storage, and habitat connectivity – are indispensable to the city's long-term resilience. However, rapid urban growth, infrastructure expansion, and changing land-use dynamics have increasingly placed these systems under stress [16, 40, 42].

Drawing on cartographic evidence from 1990 to 2018, the analysis reveals not only the persistence of high-value ecological zones but also the accelerating encroachment of urban and periurban development into riparian buffer areas (**Figure 1**).

The LULC maps depict the evolving land-use configuration of northern Istanbul, highlighting ecologically sensitive areas, watershed boundaries, and patterns of human intervention. Core hydrological features – including river corridors and water basins – form the structural backbone of the landscape, shaping both ecological connectivity and development constraints. Natural land cover (forests, wetlands, bushlands) remains concentrated in upper catchments, particularly around the Riva, Şile, and Ağva streams and the Elmalı, Ömerli, and Darlık basins.

Yet, the spatial distribution of settlements, roads, and periurban clusters shows visible signs of ecological stress. Agricultural decline is most apparent in the lower reaches of Riva and Elmalı, where fields have been transformed into periurban zones. Road infrastructure increasingly fragments forest cover, altering runoff patterns and weakening riparian buffer functions. Importantly, "Other" land cover (wetlands, bare land, water bodies, small open areas) has grown to nearly 20% of the total area by 2018, underscoring the need to better account for wetland and water body changes in planning [17].

From a planning perspective, the observed land-use patterns underscore the need for integrated watershed management strategies that balance ecological thresholds with socio-economic development. Protecting the continuity of forest and wetland systems, limiting impervious surface expansion in buffer zones, and restoring degraded riparian corridors should be central objectives. Furthermore, aligning institutional governance frameworks to address cross-boundary land-use pressures will be essential to safeguarding the ecological resilience of Istanbul's northern water landscapes in the face of climate change and ongoing metropolitan growth.

The analysis of LULC dynamics between 1990 and 2018 in Istanbul reveals significant transformations with important implications for watershed ecosystems such as Şile, Ömerli, and Ağva wetlands, which play a critical role in water regulation, biodiversity conservation, and climate resilience (**Tables 3–5**). Built-up areas within ecologically sensitive zones expanded substantially, with a modest increase of only +4.5% between 1990 and 2000, but accelerating sharply to +41% between 2000 and 2018. The annual growth rate in the latter period reached almost +2% per year, reflecting intensified pressures from speculative urbanization, second-home tourism, and infrastructure investments. This rapid encroachment directly threatens wetland buffers around Ömerli Reservoir and the riparian zones of the Riva and Ağva river systems, where construction undermines natural flood regulation and water quality. Forests remain the dominant land cover but have undergone continuous decline: -4.2% during 1990–2000 and -8.7% during 2000

LULC classes	LULC area in hectare (ha) and percentage (%)					
	1990		2000		2018	
	ha	%	ha	%	ha	%
Settlements	1,339,785	2.51	1,399,892	2.62	1,975,599	3.70
Forest	2,822,197	52.80	2,704,381	50.60	2,467,938	46.17
Agriculture	1,688,571	31.59	1,530,658	28.64	1,505,343	28.16
Bushland	6,874,628	1.29	5,005,293	0.94	9,182,129	1.72
Other*	~62,143	11.81	~91,945	17.20	~98,930	20.25

**Other: includes wetlands, water bodies, bare land, and other minor LULC classes not detailed in the table.*

Table 3.
The area coverage in hectares (ha) and percentage of LULC classes in Istanbul.

LULC Classes	1990 (ha)	2018 (ha)	Net change (ha)	Rate of change (%)	Annual rate (%/year)
Settlements	1,339,785	1,975,599	+6,358	+47.5%	+1.40
Forest	2,822,197	2,467,938	-35,426	-12.6%	-0.48
Agriculture	1,688,571	1,505,343	-18,323	-10.9%	-0.41
Bushland	6,874,628	9,182,129	+2,308	+33.6%	+1.04

Table 4.
Rate of LULC changes between 1990 and 2018 in Istanbul.

LULC Classes	1990-2000 Net (ha)	1990-2000%	1990-2000 Annual %	2000-2018 Net (ha)	2000-2018%	2000-2018 Annual %
Settlements	+601	+4.5%	+0.44	+5,757	+41.1%	+1.93
Forest	-11,782	-4.2%	-0.43	-23,644	-8.7%	-0.51
Agriculture	-15,791	-9.4%	-0.98	-2,532	-1.7%	-0.09
Bushland	-1,869	-27.2%	-3.12	+4,177	+83.4%	+3.43

Table 5.
Trend of LULC changes in Istanbul.

and 2018. The higher rate of loss in the post-2000 period coincides with the expansion of mega-infrastructure projects and the extension of transportation corridors across Istanbul's northern periphery. This decline is particularly alarming for the Ömerli basin, which relies on intact forest ecosystems to maintain groundwater recharge and reduce sedimentation in drinking water reservoirs. Agricultural areas experienced a sharp decline of -9.4% in the 1990–2000 decade, followed by a slower reduction of -1.7% between 2000–2018. The initial loss reflects the rapid conversion of arable lands into periurban residential and tourism uses, especially around Şile and Ağva, where small-scale farming once served as a buffer against land speculation. Although the decline has slowed, the fragmentation of farmland has undermined the ecological mosaic that once connected wetlands, forests, and settlement areas. Bushland cover shows a complex trajectory: a dramatic -27% reduction in the 1990s, followed by a striking +83% increase from 2000–2018. This apparent recovery is partly attributable to the abandonment of agricultural plots and transitional vegetation growth, but it also reflects ecological degradation – as degraded forests and abandoned farmlands are replaced by secondary shrublands. While this increase may suggest re-naturalization, it does not provide equivalent ecological functions compared to mature forest or wetland habitats, particularly in riparian zones of Şile and Ağva, where shrubland expansion often signals ecosystem stress rather than recovery.

The critical implications of these transformations are most evident in the Şile, Ömerli, and Ağva wetlands, which function as vital ecological buffers for Istanbul. The acceleration of settlement growth within sensitive basins has not only heightened flood risks in downstream settlements such as Ağva but also increased the pollution load that threatens the water quality of the Ömerli Reservoir. Simultaneously, the steady loss of forest cover has fragmented habitats and weakened the ecological corridors that connect wetlands to upland forest areas, directly undermining biodiversity and the resilience of Şile's coastal ecosystems. In addition, agricultural decline combined with the expansion of bushland reflects ongoing socio-economic transitions in the region. However, these changes do not necessarily enhance wetland conservation; rather, they can introduce invasive species and destabilize riparian ecosystems that are already under stress. Finally, the divergent trajectories between the 1990 and2000 and 2000 and 2018 periods highlight the absence of sustained protection mechanisms. While strategic planning documents such as the Istanbul Environmental Plan and Vision 2050 acknowledge the importance of these ecological areas, enforcement has remained weak, leaving the wetlands of Şile, Ömerli, and Ağva increasingly vulnerable to urban and infrastructural pressures.

Region	Key LULC change	Ecological implication	Governance/planning issue
Şile	Agriculture ↓, settlements ↑	Loss of buffer zones, fragmentation of wetland-forest mosaics	Weak zoning enforcement, tourism-driven development
Ömerli	Forest ↓, settlements ↑	Threats to water quality, increased sedimentation	Lack of basin-scale governance, overlapping authorities
Ağva	Agriculture ↓, bushland ↑, settlements ↑	Riparian instability, invasive species pressure, flood risk	Seasonal tourism pressure, unregulated construction
Riva	Forest ↓, agriculture ↓, settlements ↑	Habitat fragmentation along the river corridor, loss of riparian vegetation, and increased flood vulnerability	Speculative housing, inadequate enforcement of water protection zones

Table 6.
Critical ecological impacts of LULC change in wetlands.

The period 1990–2018 illustrates a clear pattern of intensified urbanization and forest loss, combined with agricultural decline and unstable shrubland expansion. These transformations have profound consequences for Istanbul's most sensitive ecological frontiers – Şile, Ömerli, and Ağva wetlands – which act as natural buffers against urban sprawl, climate change, and water insecurity. Protecting these zones requires basin-scale governance, stricter land-use control, and the integration of NBS into planning frameworks.

Table 6 synthesizes the distinct pressures on Şile, Ömerli, Ağva, and Riva. While trajectories differ, all share heightened vulnerability to urban encroachment, hydrological stress, and weak enforcement of buffer protections. The regional assessment underscores that each watershed area faces distinct but interconnected ecological pressures. In Şile, the decline of agricultural land, coupled with increased settlement expansion, has weakened the natural buffer zones protecting coastal wetlands, exposing them to fragmentation and speculative development pressures. In Ömerli, continuous forest loss and settlement encroachment pose direct threats to drinking water quality and reservoir integrity, aggravated by institutional fragmentation and lack of basin-scale coordination. Ağva illustrates a different trajectory, where agricultural decline and bushland expansion – combined with rising seasonal tourism and unregulated housing – destabilize riparian ecosystems and increase flood vulnerability. Finally, Riva emerges as a critical ecological corridor under severe pressure. The combined reduction in forest and agricultural land, together with the rapid spread of settlements, has fragmented riparian habitats and heightened flood risks along the river basin. The governance gap in enforcing water protection zones further exacerbates this ecological vulnerability. Together, these cases illustrate how diverse land-use pressures converge to undermine the resilience of Istanbul's most sensitive wetland regions.

The Ecological Risk Index (**Table 7**) provides a synthetic measure of the cumulative pressures affecting Istanbul's watershed ecosystems. Across the three decades, the index reveals a consistent increase in settlement proximity, reflecting the expansion of urbanization into once-protected buffer zones. At the same time, forest cover scores declined steadily, signaling ongoing deforestation and the weakening of critical ecological services such as water filtration, habitat connectivity, and carbon storage. Agricultural intensity decreased, but this was not

Indicator	Weight*	1990	2000	2018	Trend
Proximity to settlements	High	Low	Medium	High	
Forest cover	High	High	Medium	Medium-Low	↓
Agricultural land intensity	Medium	High	Medium	Medium	↓
Bushland fragmentation	Medium	Low	Low	Medium	
Wetland buffer integrity	High	Medium	Medium	Low	↓

Table 7.
Cumulative ecological risk index indicators in Istanbul.

necessarily ecologically positive, as abandoned farmland often shifted to fragmented bushland, which only partially substitutes lost ecosystem functions. The sharp decline in wetland buffer integrity is particularly concerning for areas like Ömerli and Riva, where wetlands provide irreplaceable services in water regulation and climate adaptation. Overall, the ERI indicates a shift from a relatively resilient system in 1990 toward a more fragmented, risk-prone landscape by 2018, emphasizing the urgent need for basin-scale governance, restoration efforts, and the integration of NBS into Istanbul's planning framework.

Figure 2 illustrates overlapping authority between ministries, Istanbul Metropolitan Municipality, and sectoral agencies. While multilayered governance has potential for collaboration, Türkiye's centralized system constrains local

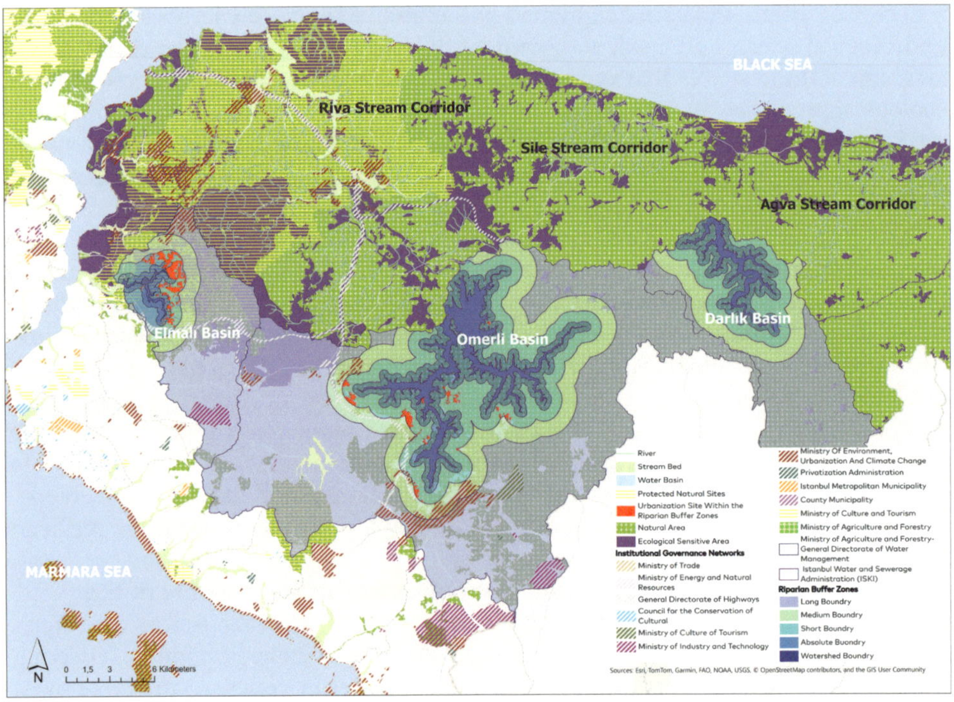

Figure 2.
Institutional governance networks (2024).

flexibility, diluting conservation outcomes. Although strategic planning frameworks such as the Istanbul Environmental Master Plan (2009) and the Istanbul Vision 2050 Action Plan (2022) acknowledge the importance of watershed protection, enforcement remains weak and fragmented [16, 17, 40].

The Institutional Governance Networks illustrate the current ecological structure and institutional governance framework of northern Istanbul, focusing on major hydrological systems such as the Riva, Şile, and Ağva stream corridors, along with the Elmalı, Ömerli, and Darlık water basins. Rivers, stream beds, and extensive riparian buffer zones are clearly delineated, with the latter classified into core, medium, and outer protection areas to safeguard water quality and biodiversity. Ecologically sensitive areas, including forests, wetlands, and designated protected natural sites, are concentrated within and around these basins, forming a critical ecological belt across the northern region. However, patterns of urbanization – particularly within riparian buffer zones – are evident, most notably along the Riva Stream corridor and near the Elmalı Basin, indicating increasing development pressure on these fragile ecosystems. The Institutional Governance Networks also reveal the overlapping jurisdiction of multiple governance bodies, including national ministries, municipal administrations, and specialized agencies, resulting in a complex network of institutional authority. While this multilayered governance structure has the potential to foster collaborative management, it also poses challenges for integrated watershed planning. The proximity of these sensitive basins to expanding metropolitan areas, combined with fragmented institutional responsibilities, underscores the urgency for ecosystem-based land-use planning that can balance urban development with the long-term protection of critical water resources.

To mitigate governance fragmentation, metropolitan and basin-level authorities could establish a Northern Watersheds Coordination Platform with legal authority to align zoning, monitoring, and enforcement across agencies. Embedding ecological thresholds into statutory plans and linking permit approvals to compliance audits would strengthen implementation. Regular joint inspections by municipal and national bodies could reduce gaps between strategic intentions and on-the-ground enforcement.

The spatial patterns revealed in the map indicate several critical risks to the region's ecological integrity and water security. First, the fragmentation of the northern forest belt, driven by infrastructure projects and dispersed urbanization, threatens the natural recharge capacity of the basins and increases vulnerability to flooding and erosion. Second, urban expansion into riparian buffer zones disrupts ecological corridors, reduces biodiversity, and compromises the self-purification capacity of water bodies. Third, the overlapping and, at times, conflicting mandates of different governance institutions hinder the timely and coordinated implementation of conservation measures. Without a unified planning approach, land-use decisions risk prioritizing short-term economic gains over long-term ecosystem health.

Beyond the aggregate land use figures, settlement density analysis by buffer zones (0–1 km, 1–3 km, 3–5 km) around the major basins and stream corridors reveals the spatial intensity of urban encroachment (**Table 8**). This perspective highlights not only the total expansion of settlements but also their proximity to critical water resources.

Table 8 demonstrates that the Elmalı and Ömerli basins are under the highest medium and outer-buffer pressures, where rural-to-periurban transformations are

Water resource/ basin	0–1 km Zone (core impact area)	1–3 km zone (medium-impact area)	3–5 km Zone (outer influence area)	Observed impact trends
Elmalı Basin	Low settlement density, scattered rural housing, and some agricultural plots	Moderate density with clustered rural neighborhoods and periurban expansion	Increasing suburbanization linked to road access	Risk of nutrient runoff from nearby agriculture; septic leakage risk in a medium buffer
Ömerli Basin	Minimal settlement, predominantly natural vegetation	Moderate density with a mixed rural-urban character; illegal construction in some areas	Expanding peri-urban zones with infrastructure corridors	Pollution risk from unregulated farming; impervious surface growth in the outer buffer
Darlık Basin	Almost no settlement; intact forest cover	Very low density, mainly seasonal-use areas	Small clustered settlements in accessible zones	Currently low pressure, but tourism growth could accelerate impacts
Riva Stream Corridor	Scattered low-density residential along riverbanks	Ribbon development along main road corridors	Expanding suburban clusters, especially toward coastal zones	High recreational use leading to bank erosion and a decline in water quality
Şile Stream Corridor	Minimal settlement, mostly forested	Low density, with scattered rural houses	Small coastal and inland villages are expanding slowly	Seasonal tourism increases waste and transport pressure
Ağva stream corridor	Minimal in core; fishing-related structures near the river mouth	Low density, small village clusters	Expansion tied to coastal tourism economy	Risk of water pollution from seasonal population spikes

Table 8
Settlement density in proximity to major water resources in Northern Istanbul.

visible. The Riva Stream Corridor, due to its scenic and recreational value, faces ribbon-type development and recreational encroachment that directly undermines riparian integrity. In contrast, Darlık, Şile, and Ağva currently exhibit lower settlement densities within core buffer zones, yet remain vulnerable to seasonal tourism influxes. These differentiated pressures highlight the importance of buffer-zone-based planning and enforcement, as low-density rural land uses are at risk of rapidly transitioning into permanent urban forms. The settlement density patterns underline the urgency of strengthening buffer-zone protections already outlined in the Istanbul Environmental Master Plan (2009) and the Istanbul Vision 2050 Strategy (2021). Without proactive governance, the gradual rural-to-urban transformation in outer buffers may irreversibly compromise the ecological resilience of Istanbul's northern basins [16, 17, 40].

As highlighted in **Table 8**, settlement densities across different buffer zones illustrate the proximity pressures on Istanbul's major basins. These patterns not only reveal where urbanization is most concentrated but also signal the varying degrees of ecological vulnerability. Building on these findings, the following section outlines policy frameworks and basin-specific assessments to address the observed challenges. To address these challenges, policy frameworks should prioritize ecosystem-based land-use planning, integrating ecological thresholds into zoning regulations and development permits. Strengthening legal protection for core and medium buffer zones, particularly in high-pressure areas such as the Riva Stream corridor and the peripheries of the Ömerli and Elmalı Basins, is essential. Inter-agency coordination mechanisms should be institutionalized, enabling the Ministry of Environment, Urbanization and Climate Change, the Ministry of Agriculture and Forestry, and local municipalities to align their planning, monitoring, and enforcement capacities. Additionally, participatory governance models – such as watershed councils involving local communities, NGOs, and academic institutions – can enhance transparency and ensure that management decisions reflect both ecological priorities and community needs. In the context of rapid metropolitan expansion, such integrated and collaborative approaches are indispensable to safeguarding Istanbul's vital water resources for future generations.

Strengthening enforcement of buffer-zone regulations requires clear responsibilities and performance indicators for each jurisdiction. A graduated sanction system – combining administrative fines, restoration obligations, and permit suspensions – could deter illegal construction. Expanding GIS-based surveillance and making violation data publicly accessible would improve transparency and compliance.

5.1 Riva stream corridor

The Riva Stream Corridor represents one of Istanbul's most critical freshwater and ecological linkages between the northern forest belt and the Black Sea. The corridor's riparian buffer zones, while still largely intact in certain upstream segments, have been increasingly fragmented by low-density housing developments, road construction, and tourism-oriented facilities. This encroachment threatens water quality, accelerates sedimentation, and disrupts the movement of wildlife species. The corridor also functions as a climate regulation zone, reducing the urban heat island effect in the surrounding metropolitan fringe. However, institutional oversight is divided between multiple agencies, resulting in limited enforcement capacity. To protect the Riva Stream Corridor, it is imperative to enforce stricter land-use controls within buffer zones, rehabilitate degraded riparian vegetation, and establish a basin-wide monitoring program supported by both municipal and community stakeholders.

5.2 Ömerli basin

The Ömerli Basin is the largest and most strategically significant drinking water reservoir on Istanbul's Asian side, supplying a substantial portion of the city's potable water. Its extensive multilayered buffer zones, as shown on the map, provide a degree of protection; however, scattered urbanization, agricultural activities, and illegal construction persist, particularly in the outer buffer zones. The basin is also vulnerable to pollution from septic systems, unregulated farming practices, and increased impervious surface coverage. Given its importance, Ömerli requires a zero-tolerance policy for new construction within core and medium buffer zones, along with targeted ecological restoration projects. Additionally, integrating spatial planning with hydrological modeling can help forecast and mitigate future risks associated with climate change, including altered precipitation patterns and drought.

5.3 Şile Stream Corridor and Darlık Basin

The Şile Stream Corridor, along with the Darlık Basin, is characterized by a relatively lower level of urbanization compared to the Riva and Ömerli catchments, yet faces growing pressure from tourism, recreational uses, and potential infrastructure expansions. The corridor supports high biodiversity, with aquatic and riparian habitats that are integral to regional ecological networks. The Darlık Basin, a significant water supply area, benefits from extensive forest cover, which acts as a natural filtration system; however, upstream disturbances, road building, and seasonal waste accumulation are emerging threats. Proactive measures should focus on maintaining the integrity of forested areas, regulating seasonal tourism activities, and enhancing community-based surveillance to prevent illegal land-use changes. The relatively intact state of these ecosystems presents an opportunity for establishing model conservation areas that demonstrate the effectiveness of integrated watershed management.

The differentiated settlement and land-use pressures become clearer when examined through the ecological assets, vulnerabilities, and policy priorities of each watershed. **Table 9** synthesizes these basin-specific dynamics, highlighting not only the risks but also the opportunities for adaptive management.

The comparative assessment highlights distinct profiles for each watershed. The Riva Stream Corridor faces the highest urbanization-driven fragmentation, requiring urgent restoration and land-use controls. The ÖmerliBasin's strategic role as a water supply area demands strict protection policies and climate-resilient management. The Şile/ Darlık system, while currently less impacted, offers a unique opportunity to serve as a demonstration site for integrated watershed conservation before development pressures intensify. A differentiated yet coordinated policy approach – tailored to each basin's ecological assets, vulnerabilities, and governance challenges – is essential to safeguard Istanbul's northern water landscapes. These three watershed areas require differentiated policy priorities – restoration in Riva, strict protection in Ömerli, and conservation-based tourism in Şile/Darlık – yet should be coordinated through a metropolitan-level watershed governance framework to ensure basin-wide resilience.

While **Table 9** provides a comparative lens, the findings converge on the need for a differentiated yet coordinated strategy across the watersheds. This is summarized in the Policy Box, which outlines concrete priorities for each basin, aligning ecological assets with targeted governance responses.

Achieving basin-wide resilience will depend on institutionalizing these differentiated strategies through a Metropolitan Watershed Governance Framework, supported by: (i) basin-specific management units with dedicated budgets; (ii) participatory watershed councils to integrate local knowledge and oversight; and (iii) legally binding conservation targets embedded in municipal development plans.

In addition, creating an independent environmental ombudsman or specialized "green court" could provide a clear avenue for addressing violations and ensuring that enforcement decisions are transparent, timely, and free from political interference.

DOI: http://dx.doi.org/10.5772/intechopen.1014148

Watershed/ corridor	Ecological assets	Current vulnerabilities	Potential for improvement	External pressures	Implications/policy priorities
Riva Stream Corridor	Continuous hydrological link from forest belt to Black Sea • Biodiversity-rich riparian vegetation • Climate regulation for urban fringe	Fragmented buffer zones due to housing and roads • Tourism-related pollution • Weak inter-agency enforcement	• Riparian restoration programs • Basin-wide participatory monitoring	• Accelerating suburban sprawl • Infrastructure expansion projects	Priority: Riparian habitat restoration + strict tourism management. Establish integrated corridor management involving local municipalities to enforce land-use controls.
Ömerli Basin	• Largest drinking water source on the Asian side • Multilayered legal protection zones • Strategic role in metropolitan water security	• Scattered illegal construction • Agricultural runoff and septic leakage • Impervious surface	• Zero-tolerance policy for core buffers • Climate-adaptive hydrological planning	• Drought and climate variability • Population-driven demand growth	Priority: Absolute protection of core zones + climate-resilient water management plan. Strengthen enforcement via the watershed council to align local and national institutions' responsibilities
Şile Stream Corridor and Darlık Basin	• High biodiversity with intact forest cover • Lower urbanization pressure Natural filtration by forest ecosystems	• Seasonal waste accumulation • Road building in upper catchments	• Model conservation area potential • Regulated eco-tourism development	• Tourism boom without controls • Infrastructure encroachment	Priority: Conservation-led eco-tourism model with a seasonal waste management plan. Designate as a demonstration site for community-based forest and stream protection

Table 9.
Comparative analysis of key watershed areas in northern Istanbul.

Policy Box: Differentiated but coordinated strategies
Riva Corridor → Riparian restoration + strict control of tourism and housing.
Ömerli Basin → Absolute protection of core/medium buffers + climate-resilient hydrological planning.
Şile and Ağva Corridors → Eco-tourism model with regulated seasonal use + community-based monitoring.
Darlık Basin → Designated as a model conservation site to demonstrate proactive integrated management.

6. Conclusion

This chapter critically examines Istanbul's northern watershed regions – Riva, Ömerli, and Şile – through an integrated framework combining LULC analysis, ecological risk assessment, and governance mapping. The results reveal a growing mismatch between ecological thresholds and planning outcomes, illustrating how unregulated urban expansion, fragmented governance, and weak enforcement mechanisms undermine long-term water security and ecosystem resilience. These findings resonate with broader evidence from rapidly urbanizing regions, where uncoordinated land conversion has been shown to intensify ecological risks and accelerate biodiversity decline [7, 8].

Moreover, the LULC analysis (**Tables 3–5**) demonstrates a striking pattern of intensified settlement growth and forest decline over the past three decades. Built-up areas expanded by + 47.5% between 1990 and 2018, with the fastest increase observed after 2000, coinciding with speculative housing and infrastructure investments. Forests, the primary ecological backbone of the basins, declined by 12.6%, fragmenting habitat corridors and reducing recharge capacity, particularly in the Ömerli basin. Agricultural areas shrank by 10.9%, while bushland increased irregularly, reflecting ecological degradation rather than true regeneration. These transformations mirror global studies indicating that unchecked peri-urban growth and forest loss directly weaken watershed resilience and heighten climate-related risks [9, 13]. However, the Ecological Risk Index (**Table 7**) further underscores these vulnerabilities, with settlement proximity shifting from low (1990) to high (2018), and wetland buffer integrity falling from medium to low. These changes signify heightened flood risk, declining water quality, and reduced biodiversity. Site-specific pressures reinforce this trend: Ribbon development in the Riva corridor, gated community projects in Ömerli, and coastal tourism pressures in Şile all demonstrate how localized land-use decisions collectively escalate metropolitan-scale ecological risks. Similar patterns have been observed in other Global South contexts, where ecological thresholds are routinely exceeded due to weak policy enforcement and fragmented governance [20, 22].

Furthermore, the institutional analysis highlights systemic weaknesses that exacerbate ecological pressures. The Policy Gap Analysis (**Tables 1–2**) reveals that while regulations such as the Water Pollution Control Regulation (2004) and Drinking Water Basin Protection Regulation (1998/2004) provide strong legal bases, enforcement is inconsistent. For example, buffer zones in Riva have been rezoned for residential and tourism use despite explicit prohibitions. Strategic plans such as the Istanbul Environmental Master Plan (2009) and Vision 2050 acknowledge ecological priorities but remain aspirational due to a lack of binding enforcement, dedicated funding, and performance indicators. These governance failures echo broader critiques that emphasize the persistence of sectoral, engineering-led approaches over ecosystem-based planning, undermining resilience and perpetuating reactive rather than preventive strategies [4–6].

The comparative framework (**Table 9**) highlights both shared risks and differentiated vulnerabilities:

- Riva Stream Corridor → Severe riparian fragmentation, weak enforcement, and suburban sprawl; urgent need for riparian restoration and strict tourism management.

- Ömerli Basin → A strategic drinking water source, yet highly vulnerable to illegal construction and agricultural runoff; it requires zero-tolerance buffer enforcement and climate-adaptive hydrological planning.

- Şile and Darlık Corridors → Currently less urbanized but threatened by seasonal tourism and waste accumulation; potential to serve as a model conservation site with eco-tourism regulations.

On the other hand, these findings converge on a common conclusion: while each watershed requires tailored strategies, they must be embedded within a coordinated metropolitan-level watershed governance framework. This chapter makes two key contributions. First, it provides the first integrated LULC + governance risk framework for Istanbul's watersheds, combining spatial metrics (fragmentation index, settlement density, ERI) with institutional and policy analysis. Second, it demonstrates how EBLUP can be operationalized in contexts marked by strong centralization, policy fluidity, and conflicting sectoral mandates – issues characteristic of the Global South but underexplored in the literature [1, 2].

However, to address identified risks, the following strategic interventions are recommended: Riparian buffer restoration to enhance flood retention and biodiversity; strict zoning enforcement in drinking water basins, closing regulatory loopholes; integration of NBS into statutory plans, mandating wetlands, forests, and corridors as functional infrastructure; participatory watershed councils to align central, municipal, and community stakeholders; and GIS-based monitoring platforms for real-time detection of settlement encroachment and ecological stress. These measures not only mitigate ecological risks but also align Istanbul's planning system with global sustainability agendas, particularly SDG 6 (Clean Water), SDG 13 (Climate Action), and SDG 15 (Life on Land) [19].

The findings emphasize that protecting Istanbul's northern watersheds requires policies that go beyond technical planning. Stronger enforcement of buffer-zone regulations, transparent monitoring systems, and clear accountability for agencies are essential to prevent irreversible ecological loss. Embedding ecological thresholds and climate resilience targets into statutory plans would help align day-to-day development decisions with long-term sustainability goals.

For future research, there is scope to develop scenario-based models that combine land-use change, ecological thresholds, and governance responses to test how different planning and enforcement choices affect water security and biodiversity.

Long-term datasets on compliance and restoration outcomes could help assess which interventions deliver the most durable results. Comparative analyses with other rapidly urbanizing watersheds would extend the applicability of the framework presented here, revealing how institutional design and participatory approaches shape ecological resilience under different political and socio-economic settings.

Together, these directions underline that sustaining metropolitan water landscapes depends on coupling robust spatial evidence with adaptive and accountable institutions, ensuring that conservation and development objectives reinforce, rather than undermine, one another.

The evidence presented here shows that Istanbul's northern waterscapes are at a critical tipping point. Without decisive reform, ecological thresholds may be irreversibly crossed, undermining both urban water security and climate resilience. By linking spatial transformations with governance failures, this chapter offers a replicable framework for integrating EBLUP and NBS into metropolitan planning, contributing both academically and practically to the rethinking of land-use governance in rapidly urbanizing watershed regions.

Conflict of Interest

The authors declare no conflict of interest.

Author details

Ozge Celik Yilmaz

1 Istanbul Technical University, Istanbul, Turkiye

2 Faculty of Architecture, Taşkışla Campus, Elmadağ, Istanbul Technical University, Istanbul, Türkiye

*Address all correspondence to: ozgcelik@itu.edu.tr

References

[1] Marchioni M, Raimondi F, Becciu G, Dresti C. Nature-based solutions for watershed management: An investigation on water-related ecosystem services delivery at multiple spatial scales. Ecosystem Services. 2025;**73**. DOI: 10.1016/j.ecoser.2025.101718

[2] Wamsler C, Wickenberg B, Hanson H, Alkan Olsson J, Stålhammar S, Bj€orn H, Gerell D, Oskarsson T, Simonsson E, Torffvit F, Zelmerlow F. Environmental and climate policy integration: Targeted strategies for overcoming barriers to nature-based solutions and climate change adaptation. Journal of Cleaner Production. 2020;**247**. DOI: 10.1016/j.jclepro.2019.119154

[3] Dem P, Hayashi K, Fujii M, Mohmmed EEE. Integrating biodiversity and ecosystem services in land use change assessment through sustainability indicator. Environmental Impact Assessment Review. 2025;**114**. DOI: 10.1016/j.eiar.2025.107971

[4] Healey P. Collaborative Planning: Shaping Places in Fragmented Societies. 2nd ed. UK: Palgrave Macmillan; 2006

[5] Pahl-Wostl C. An evolutionary perspective on water governance: From understanding to transformation. Water Resources Management. 2017;**31** (10):2917–2932. DOI: 10.1007/s11269-017-1727-1

[6] Liu B, Wang M, Chen T, Tseng P, Sun Y, Chiang A, Chiang P. Establishment and implementation of green infrastructure practice for healthy watershed management: Challenges and perspectives. Water-Energy Nexus. 2020;**3**:186–197. DOI: 10.1016/j.wen.2020.05.003

[7] Malede DA, Alamirew T, Kosgie JR, Andualem TG. Analysis of land use/land cover change trends over Birr River Watershed Abbay Basin, Ethiopia. Environmental and Sustainability Indicators. 2023;**17**. DOI: 10.1016/j.indic.2022.100222

[8] Wei Y, Zhou P, Zhang L, Zhang Y. Spatio-temporal evolution analysis of land use change and landscape ecological risks in rapidly urbanizing areas based on Multi-Situation simulation - a case study of Chengdu Plain. Ecological Indicators. 2024;**166**:112245. DOI: 10.1016/j.ecolind.2024.112245

[9] Jat RA, Jinger D, Kumawat A, Kar S, Rawat I, Kumar S, Paramesh V, Meena VS, Kaushal R, Kumar K, Meena HS, Wani SP, Singh R, Madhu M. Integrated watershed management for transforming dryland livelihoods: A climate-smart strategy for sustainable dryland agriculture in India. Watershed Ecology and the Environment. 2025;7:159–177. DOI: 10.1016/j.wsee.2025.03.006

[10] Qin C, Zhu L, Zhu A. Position paper: Domain knowledge-driven intelligentization of watershed modeling and scenario analysis for precise watershed management. Information Geography. 2025;**1**:100016. DOI: 10.1016/j.infgeo.2025.100016

[11] Worku G, Teferi E, Bantider A, Dile YT. Prioritization of watershed management scenarios under climate change in the Jemma sub-basin of the Upper Blue Nile Basin, Ethiopia. Journal

of Hydrology: Regional Studies. 2020;**31**. DOI: 10.1016/j.ejrh.2020.100714

[12] Kaufman DE, Shenk GW, Bhatt G, Asplen KW, Devereux OH, Rigelman JR, Ellis H, Hobbs BF, Bosch DJ, van Houtven GL, McGarity AE, Linker LC, Ball WP. Supporting cost-effective watershed management strategies for Chesapeake Bay using a modeling and optimization framework. Environmental Modelling and Software. 2021;**144**:105141. DOI: 10.1016/j.envsoft.2021.105141

[13] Wolka K, Uma T, Tofu DA. The role of integrated watershed management in climate change adaptation for small-scale farmers in Southwest Ethiopia. Environmental and Sustainability Indicators. 2023;**19**:100260. DOI: 10.1016/j.indic.2023.100260

[14] Fenta AA, Tsunekawa A, Haregeweyn N, Tsubo M, Yasuda H, Kawai T, Berihun ML, Ebabu K, Sultan D, Mekuriaw S. An integrated framework for improving watershed management planning. Environmental Research. 2023;**236**. DOI: 10.1016/j.envres.2023.116872

[15] Derib G, Alemayehu A. Assessment of sustainability in the Wulo Abiye watershed, central highlands of Ethiopia. Environmental Challenges. 2024;**15**:100934. DOI: 10.1016/j.envc.2024.100934

[16] Istanbul Metropolitan Municipality (IMM). Istanbul Vision 2050 Strategic Document. Istanbul, Türkiye: Istanbul Planning Agency; 2021

[17] Republic of Turkiye Ministry of Environment, Urbanization and Climate Change. Istanbul Environmental Plan (1/100.000 scale); 2019

[18] Tournebize J, Vincent B, Chaumont C, Gramaglia C, Margoum C, Molle P, Carluer N, Gril J. Ecological services of artificial wetland for pesticide mitigation Socio-technical adaptation for watershed management through TRUSTEA project feedback. Procedia Environmental Sciences. 2011;**9**:183–190. DOI: 10.1016/j.proenv.2011.11.028

[19] IPCC. Climate Change 2022: Impacts, Adaptation, and Vulnerability. LONDON, UK: Cambridge University Press; 2022

[20] Zhang M, Gu W, Yang D, Ruan Y. Influence of multilevel governance on land use change in China's rapidly urbanizing metropolitan from low-carbon perspective. Environmental and Sustainability Indicators. 2024;**23**:100450. DOI: 10.1016/j.indic.2024.100450

[21] Wahdatyar R, Khokhar MF, Ahmad S, Rahil MU, Stanikzai MA, Khan JA, Kamran. Exploring the dynamics and future projections of land use land cover changes by exploiting geospatial techniques; A case study of the Kabul River Basin. Heliyon. 2024;**10**:e39020. DOI: 10.1016/j.heliyon.2024.e39020

[22] Kavhu B, Mashimbye ZE, Luvuno L, Sanga U. Integrating socio-economic zones into water resource analysis under land use change and climate variability in the Okavango basin, Watershed. Ecology and the Environment. 2025. DOI: 10.1016/j.wsee.2025.06.003

[23] Jamilah M, Becker A, Lee SL, Quiros A, Chee SY, Evans C, Tan YY,

Ti LL, Pishal I, Then A. Socioeconomic impacts linked to land use and land use changes affecting blue carbon ecosystems in Southeast Asia: A systematic map. Ocean and Coastal Management. 2025;**267**. DOI: 10.1016/j.ocecoaman.2025.107643

[24] Smit I, Maze K, Wilgen BW. Land cover change in and around South African protected areas. Biological Conservation. 2024;**300**:110844. DOI: 10.1016/j.biocon.2024.110844

[25] Adarkwah F, Awuni S, Hajek M, Kübler D, Mattah M, Gordon C, Owusu E. Modelling the drivers of land use and land cover change of the great Amanzule wetland ecosystem to inform the development policy of the southwestern oil-rich region of Ghana. Heliyon. 2024;**10**:e36635. DOI: 10.1016/j.heliyon.2024.e36635

[26] Wittberg E, Tavares AF, Szmigiel-Rawska K. Local politics and land take: Using remote sensing data to analyse land-use changes in Sweden. Journal of Environmental Management. 2025;**384**:125387. DOI: 10.1016/j.jenvman.2025.125387

[27] Kassun BW, Kallio M, Trømborg E, Rannestad MM. Land use and land cover change, trade-offs, and synergies between ecosystem services in a dry Afromontane Forest. Journal for Nature Conservation. 2025;**85**. DOI: 10.1016/j.jnc.2025.126874

[28] Shuangmi X, Wenzhu Z, Bin W, Shengquan L, Guanhai G, Yanhua L. Evaluating the migration of boundary river shorelines and coastal land cover changes for the Beilun River between China and Vietnam. Journal of Hydrology: Regional Studies. 2025;**57**. DOI: 10.1016/j.ejrh.2024.102167

[29] Zaman W, Real HRK. Machine learning-based spatio-temporal assessment of land use/land cover change in Barishal district of Bangladesh between 1988 and 2024. Environmental Challenges. 2025;**19**. DOI: 10.1016/j.envc.2025.101168

[30] Wang Q, Bai X, Zhang D, Wang H. Spatiotemporal characteristics and multi-scenario simulation of land use change and ecological security in the mountainous areas: Implications for supporting sustainable land management and ecological planning. Sustainable Futures. 2024;**8**:100286. DOI: 10.1016/j.sftr.2024.100286

[31] Wang L, Liu X, Lei J, Ma J, Zhang L, Liu X. How the land use change impact on ecological vulnerability in the Lanzhou-Baiyin section of the Yellow River Basin, China. Environmental and Sustainability Indicators. 2025;**27**:100779. DOI: 10.1016/j.indic.2025.100779

[32] Miralles-Wilhelm F, Matthews JH, Karres N, Abell R, Dalton J, Kang S, Liu J, Maendly R, Matthews N, McDonald R, Munoz-Castillo R, Ochoa-Tocachi BF, Pradhan N, Rodriguez D, Vigerstøl K, van Wesenbeeck B. Emerging themes and future directions in watershed resilience research. Water Security. 2023;**18**:100132. DOI: 10.1016/j.wasec.2022.100132

[33] Ohnemus T, Dirnböck T, Bäck J, Gaube V, Kühn I, Mirtl M, Mollenhauer H, Vereecken H, Zacharias S. Fitness for future: ELTER RI's representation of climate and land use change. Ecological Indicators. 2025;**171**:113159. DOI: 10.1016/j.ecolind.2025.113159

[34] Liu R, Wang Q, Kong H, Li Y. Priority protected areas for mangrove conservation in coastal Guangdong, China: Addressing climate and land cover changes. Ocean and Coastal Management. 2025;**267**:107707. DOI: 10.1016/j.ocecoaman.2025.107707

[35] Lu J, Calabria J, Hawkins G. Exploring potential mechanisms linking urban landscapes to instream stressors and macroinvertebrates in the Southeast Piedmont: Implications for watershed management. Ecological Indicators. 2024;**163**:112114. DOI: 10.1016/j.ecolind.2024.112114

[36] Li H, Zhang S, Zhang J, Zhang W, Song Z, Yu P, Xie C. A framework for identifying priority areas through integrated eco-environmental risk assessment for a holistic watershed management approach. Ecological Indicators. 2023;**146**:109919. DOI: 10.1016/j.ecolind.2023.109919

[37] Yilmaz O, Ertekin O.Urban living labs as a tool to achieve sustainable development goal 16: A case study of Istanbul, Turkiye. Journal of Theoretical and Empirical Researches in Urban Management. 2023;**12**(2):45–62

[38] TUBITAK-MAM. Climate Change Impacts on Istanbul's Water Resources. ANKARA, Türkiye: Marmara Research Center; 2018

[39] IWSA. Riva Basin Protection and Water Quality Monitoring Report; 2021; Istanbul Water and Sewerage Administration.

[40] Istanbul Metropolitan Municipality (IMM). Istanbul Vision 2050 Action Plan. Istanbul, Türkiye: Istanbul Planning Agency; 2022

[41] IWSA. Annual Water Quality Status Report of Omerli Reservoir; 2022; Istanbul Water and Sewerage Administration.

[42] Istanbul Metropolitan Municipality (IMM). Istanbul Green City Action Plan; 2025

Chapter 4

Retrospective and Prospective Views on the Dynamics and Governance of the Integrated and Nested Nature and Society Ecological Complexes (NSEC): The Eastern Lower Danube River Sub-catchment

Angheluță Vădineanu, Anca Crăciun and Radu Ștefan Vădineanu

Abstract

Considering land and open water use planning, and adaptive and transformative governance for a sustainable future as the most challenging and wicked thematic area, we propose a holistic approach to address and draw on: dynamics of the **Earth Life Supporting Environment (ELSE)** across spatial and temporal scales and the position and impact of human societies within it. We develop conceptual and analytical frameworks that enable identification and interpretation of structural and functional complexity for planning and governance of the sustainable transition of nested Nature -Society Ecological Complexes (NSEC). The Romanian Lower Danube River sub-catchment (RoLDRsC), including the river and floodplain wetlands (Ro-LDW), is used as a test case. The chapter integrates historical contextualization, theoretical developments from systems ecology and sustainability science, and empirical long-term socio-ecological research to clarify critical questions regarding sustainable transition: when it occurs, why it emerges, which structural transformations are involved, and what spatial and temporal scales. We examine major land-use transformations, governance regimes, and socio-economic transitions, and assess their ecological trade-offs and systemic implications. Particular attention is given to structural thresholds, ecosystem services dynamics, and resilience within the Lower Danube wetlands. **Finally, the chapter advances a normative scenario framework to support adaptive and transformative governance**, proposing pathways for long-term sustainable transition of nested socio-ecological systems under conditions of uncertainty, institutional change and evolving societal demands.

IntechOpen

Keywords: natural capital, thresholds, tipping points, conservation, restoration, trade-offs, sustainability

1. Introduction

The second half of the 20th century has been recorded as a time of **great acceleration** in economic and demographic growth, along with increasing public and policy and decision makers concerns regarding the high levels of human-induced **environmental degradation.**

It was also the time when two opposed concepts – **crisis** and **sustainability** for describing human and environment relations – had been widely and intensively addressed in public and policy discourses. In the meantime, the formally established integrative science of **Systems Ecology** contributed to the clarification and improvement of the concept's definition, which helped significantly in the preparatory activities and documents (Brundtland Report/1987 for Agenda 21) for the first UNCED-Rio1992.

Since then, the global and decades-long process of **sustainable transition (ST)** toward a generic, yet poorly defined, **vision** or model of socio-economic sustainability in conjunction with the natural environment has been launched.

The leading team authors of this paper have been actively involved, since the preparatory phase along with the three following phases of **global sustainable transition (GST),** at different levels (and positions), but always as academic teachers and researchers at the University of Bucharest. Combining their expertise and that of the other two co-authors, as well as those of many national and international collaborators, obtained from critical analysis and synthesis of a large and diverse number of sources – scientific and gray literature, proceedings, technical reports, and very valuable insights from open debates with different stakeholders, including students, academics, policy and decision-makers, and managers of private and public institutions – we decided to focus this contribution on the most challenging conceptual and operational conditionalities required for better system identification and understanding of structural and functional complexity, across space and time, of the **Earth life support environment (ELSE)** for effective planning and adaptive governance of the "**global sustainable transition (GST).**"

Thus, we make a clear distinction between two prototypes of GST. First, the sustainable transition of socio-economic systems targeted for **continuous economic growth** and **human well-being,** with increasing consideration of the impact on the structural and functional capacity of the **natural environment** /or **natural capital** [1–6]. This type of sustainable transition is guided by a dynamic 3D (social, economic, environmental/ecological) conceptual framework.

Sustainable socio-economic development requires: a) continuous improvement of the knowledge concerning the retrospective dynamics of the complex biotic and abiotic relationships, including those between human societies and the **natural environment** during the last geological epoch – Holocene, for understanding the roots and drivers of the **ecological crisis**; b) theoretical and conceptual developments aiming for system identification and hierarchical (nested) organization as social-ecological systems across space and time scales of the Earth's environment; c) operational infrastructure combining political programs, strategies and action plans, legal, fiscal and institutional arrangements, research and decision support platforms, advanced cyber and communication technologies, procedures, methods, and tools; d) extensive **education about** and **training for** effective participation and contribution, on large space and time scales, adaptive and transformative ST governance of particular or a set of nested 3D – SES.

The co-occurrence at certain levels of these conditionalities defines a particular **governance regime** applied to transformative changes [7, 8]. A similar approach has been extensively used in many studies focused on land-based ecosystems [9] to distinguish between **land use change** and **land management regimes.** The land was a cornerstone of human civilization and is currently linked to many global sustainability challenges – from climate change to food security and biodiversity loss, as well as to new control variables for **functional biodiversity integrity** within the planetary boundaries [10]. A large recognition that contemporary landscapes are contingent outcomes of past and present patterns, processes, and decisions indicates that empirical analysis of past and present land use change has an important role in providing insights into the socio-economic and ecological processes that shape land use transitions and land management regimes [11]. Changes in socio-ecological metabolism, land use change, and changes in governance and communication processes have been identified as major research themes in LTSER [12].

Second, a recently emerging prototype of sustainable transition refers to the sustainable transition of the **Earth' ecosphere** and its fully **integrated** and **nested natural and social environments** ($hNSEC^s$), down to local ones.

We underpin the proposed holistic prototype on the most recent and advanced theoretical and conceptual acquisition by the fast-emerging **transdisciplinary** and **holistic science** – Systems Ecology of Sustainability, which helps extend holistic system identification and understanding of structural and functional complexity across space and time within the Earth's life support environment.

The above critical conditionalities and types of sustainable transition receive **synthetic and state-of-the-art** conceptual and theoretical support in sections 2 and 3, while the sections 4, 5, 6 and 7 provides a synthesis of the approaches and findings of 3D types of sustainable transition in one of the most dynamic and interconnected sets of nested socio-ecological systems, as well as practical support to shift to the second type of holistic sustainable transition.

2. Historical contextualization of the sustainable transition (ST)

2.1 Biosphere and ecosphere: Life and environment life support

The emergence and explosive development of unicellular prokaryotes, represented by photosynthesizing blue-green algae about a billion years ago, was the source of profound changes in the composition, structure, and role of the atmosphere and hydrosphere, together with the role of the primary geosphere in the configuration and functioning of the Earth system as a support system for the development and evolution of the Biosphere [13–19].

Over a period of approximately 600 million years, a series of relatively cyclical transformations have taken place in the composition, structure, and activity of the terrestrial and aquatic components of the biosphere, which have marked the role of this component in supporting, orienting, and controlling the dynamics and evolution of the Earth Life Support System (Ecosphere) (**Figure 1**).

For hundreds of thousands of years (the last part of the Pleistocene, Quaternary period) [13], the human species, Homo sapiens, has been integrated into and adapted to the structure and functioning regime of the Biosphere, whose contribution to the

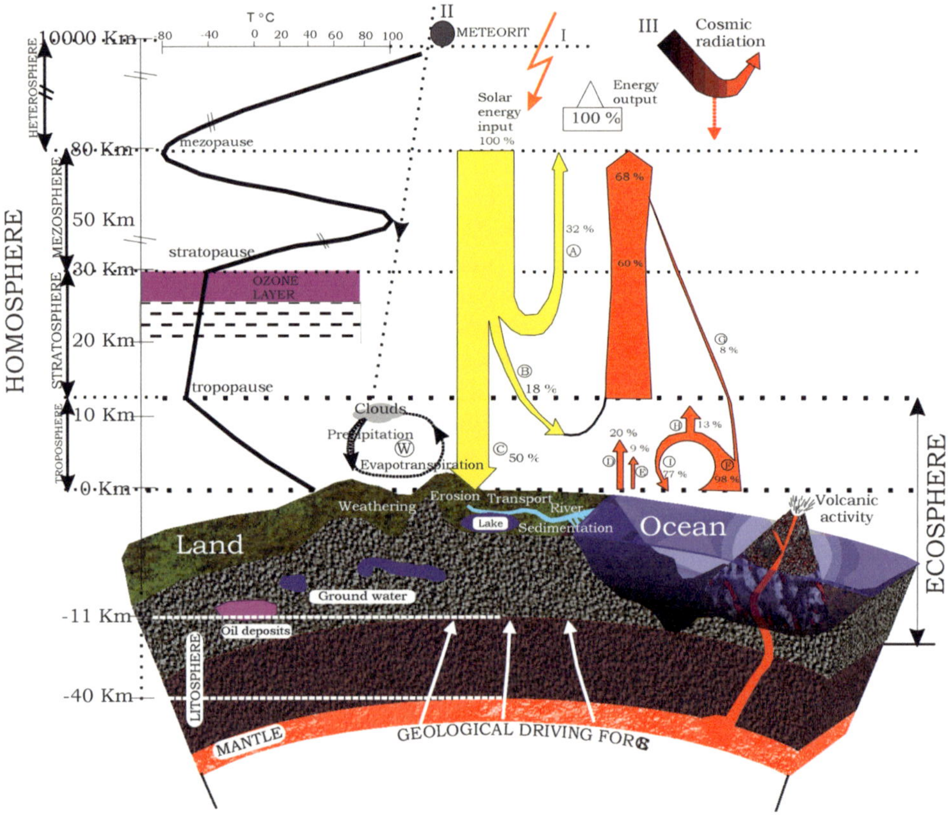

Figure 1.
Space distribution of the ecosphere within the Earth-atmosphere system and the energy balance (compiled by Vadineanu, 1998, 2001, from *Strahler* et al., *1974 and D. Chiras, 1991). Total solar energy input (Es = 100%): (a) Reflected energy by lower atmospheric layers (32%); (b) Energy absorbed by the atmospheric layers (18%); (c) Incident energy at land and ocean surface (50%). Total radiated energy by the ecosphere (68%): (d) Latent heat (evapotranspiration) – 20%; (e) Conduction and convection (kinetic energy – 9%); (f) Radiated energy by land and ocean surface (98%); (g) Direct radiation to space from land and ocean surface (8%); (h) Radiation from troposphere to space (13%). Energy output back to space (100%), consisting of reflected (32%) and radiated (68%) energy by the ecosphere; (i) Contraradiation from troposphere back to land and ocean surface – 77% (greenhouse loop).*

adaptive transformations and evolution of the Ecosphere during the Pleistocene was dominant or at least equivalent to that of the other components of the Earth system [18].

Recently, it has been accepted that in the last geological epoch – the Holocene (about 12 thousand years after the last big glaciation), the biosphere had a major contribution to setting up the stability domain and resilience of the Earth's ecosphere [13].

Thus, favorable conditions have been created for the human species (modern, rational man), Homo sapiens, to transform and use the biophysical structure of "nature" [20].

2.2 Emergence of the ecological crisis

Retrospective studies have analyzed the dynamics of these relationships since the mid-18th century and have highlighted a gradual increase in the number, intensity, and cumulative effects of the pressure of human societies (social and economic systems) on the composition, structure, and functional capacity of the Biosphere as a **global life system** and the Ecosphere as a global **Environment Life Support System**, which would orient their dynamics toward a geological epoch dominated by instability, multiple crises, and a global **ecological crisis** [7, 21–26].

Crutzen P.J., Vice-President of the International Geosphere-Biosphere Program (IGBP) and member of the IGBP project team (1999–2003), which aimed at the analysis and synthesis of the results of the IGBP's long-term research sectoral projects, proposed the concept of "Anthropocene" to define the new geological epoch and suggested that the transition from the Holocene to this new epoch, induced and sustained by the increasing pressure of "human activities," would have been triggered in the second half of the eighteenth century, with the beginning of the Industrial Revolution [27, 28].

Homo sapiens has transformed and acted over the last three centuries as the main dominating and destabilizing force at the level of the biosphere and the ecosphere, with climate, social, and economic implications at a global level (**Figure 2**)

Aggregation, analysis, and interpretation of the large volume of primary data and information generated by: i) the implementation of research and monitoring programs and projects under the auspices of UNESCO and ICSU in the second half of the twentieth century, focused on – geological resources (IGSP); water resources (IHP); biological resources (DIVERSITAS); interactions between the geosphere and biosphere components of the Earth System (IGBP); Climate and Climatic System (WCRP); relations between socio-economic systems and the biosphere (MAB); and the management of social transformations (MOST) [7, 24, 26, 29]; ii) databases and reports of international agencies such as UN-FAO, WMO, and WHO – total and urban population; iii) reports of the secretariat of the Intergovernmental Organization on Climate Change (IPCC) and the Intergovernmental Platform on Biodiversity and Ecosystem Services (IPBES); iv) banking and financial institutions;

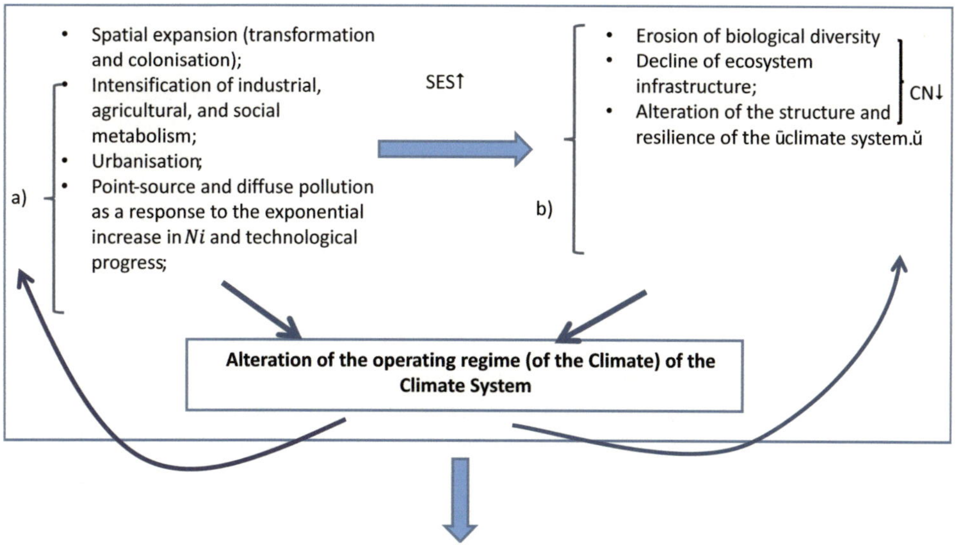

Figure 2.
"Crisis" in the relationships between socio-economic systems and the natural environment (NC), or between economic and social development and the natural environment, or between the social environment and the natural environment.

v) national, regional, and global statistical centers or institutes, etc. [30–37], made it possible to highlight trends in the dynamics of the set of 12 indicators specific to the Global Socio-Economic System, and differentiated for OECD member countries, the group of countries with emerging economies (BRICS), and the rest of the underdeveloped countries, respectively, the set of 12 indicators specific to the "Earth Life Support System," during the period 1750 - 2010 [38].

Most studies that have focused on the dynamics of the relationships between "people and the environment," specifically between human societies and biological and ecological systems, have made an arbitrary delimitation between "social space" – with demographic, economic, political, technological, and cultural dimensions – and "natural space" within the Earth System Ecosphere [1, 7, 39]. (**Figure 2**)

Such an interpretation of the society and environment relations suggests a complete overlap of the environment and nature positions within the Earth system's structure. According to this perception, the environment/nature concepts have been defined and used in academic, policy, and public discourses, without any clear specification, as a set of environment/nature factors (e.g., air, water, land/or soil, vegetation). The related issues have been identified and managed based on sectoral and reductionist principles, which have not stopped the accelerated dynamics of the global ecological crisis.

However, we noticed that in the last 2–3 decades, parallel but significantly improved conceptual frameworks have been developed and applied for system and hierarchical identification and for understanding the organization of the Earth environment across space and time scales. This big shift upward, from ambiguity and sectoral identification of the environment as nature outside **social space, first to 3D-integrated social and natural Earth environment systems (3D-SES)**, was possible due to the great theoretical achievements of two level 1 integrative sciences: **a) Systems Ecology** and **b) Sustainable Development**, and more recently to holistic Nature and Social Ecological Complexes (h-NSEC^s) based on theoretical and operational achievements in the emerging level 2, Trans disciplinary Science: Systems Ecology of Sustainability of h-NSEC^s.

The consequences of such approaches and interpretations on the position, relations, and development of human societies, especially in the last three centuries, have supported the conclusions according to which the development of human societies has been achieved at the expense of nature and the climate system over large geographical areas [8, 40, 41]. **Figure 3**.

2.3 Sustainable transition (ST) (when?)

Significant progress has been made in understanding and arguing that global human society development (patterns and rates) has become, since 1950, the major driving force of global environmental changes. The process has been defined as a fast-widening dichotomy between socio-economic development (D) and the quality deterioration of the Earth's environment (E), often described as the "global ecological crisis" [5, 42], as well as a transition from the Holocene World (named and frequently referred to as the normal World) [43], with relative stability based on biosphere and ecosphere resilience, to a new unpredictable post-normal World driven by the human-created and less resilient **Anthroposphere**. The current emerging "world" is usually described in the scientific literature as the new geological epoch – **Anthropocene** [28, 38].

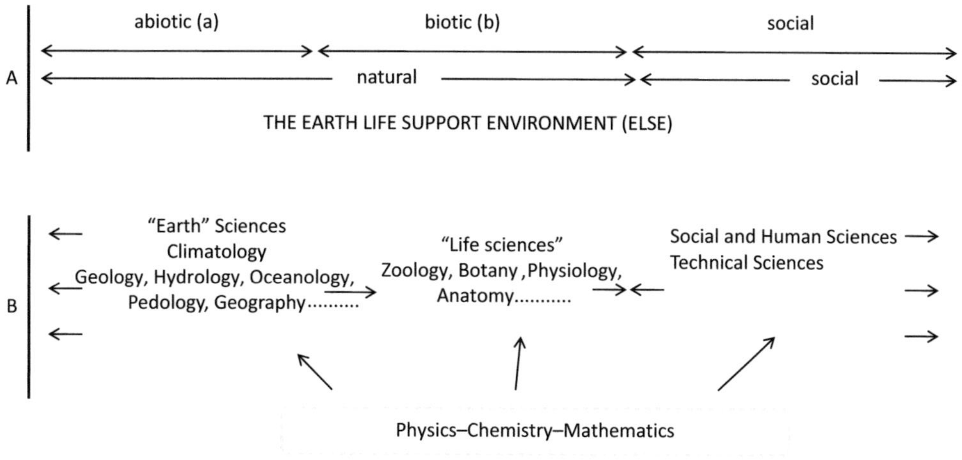

Figure 3.
Spectrum of (a) the Earth environment and (b) the scientific domains and subdomains that have contributed to the processes of intra-, inter-, and trans-disciplinary integration, and thus to the emergence and development of systems ecology.

The response to such an unpredictable future for humanity has been designed and launched as a long-term sustainable transition toward a balanced and adaptive new world in a safe, life-supporting environment/or **safe Ecosphere.**

As such, the long-lasting sustainable transition proves to be more challenging in each of the two decadal phases – (post-Rio/1992, post-Johannesburg/2002) and the current extended phase (post Rio 20 +, UN Agenda 2030/2015).

Along past and upcoming phases of sustainable transition, challenges and wicked issues related to land cover, land use planning, and continental and coastal water management and intensification have been identified as among the most powerful stressors shaping trends and outcomes.

The intensification of agriculture has been a general trend in Europe since the end of the Second World War, generating homogenization of the landscape [11].

The lowlands of South – East Romania, extending over 75% of the total surface of the Lower Danube Sub-catchment (Ro-LDR$_s$C), represent a typical example (like the entire surface of arable land of nine million hectares on Romanian territory) [National Institute of Statistics/NIS, 2003] of such phenomenon: very large (>1,000 ha) of crop fields and disturbed or lost habitats – such as hedges, groves, and woods.

These types of large and intensive agricultural ecosystems show a high spatial homogeneity, a low species richness and genetic diversity, and highly disturbed food webs, all features that contribute to a decrease in biodiversity and Eco diversity in the Ro-LDR$_s$C [44].

3. Theoretical and conceptual background

The post-normal world, which gradually emerged since the 18th century [28], followed by the **accelerated transition** since the mid-20th century [28] toward a new geological epoch – Anthropocene mostly driven by social and economic stressors, is characterized by **complexity, chaos, and contradictions** [43, 45–47].

In such circumstances, "**nothing is definite, truly guaranteed or total safe**" [48], and disciplinary or even less integrated interdisciplinary knowledge, theories, concepts, and sectoral approaches, with their specific tools, do not help to address complexity and the related wicked problems aimed at preventing the real danger of a highly integrated **ecosphere** running out of Earth's life-supporting boundaries [49–51].

3.1 Level 1 integrative sciences

First attempt to address emerging scientific theories and operational challenges concerning humanity-driven transition of the Ecosphere to a **post-normal World/ Anthropocene** was the proposal for interdisciplinarity and integrated "science" for the post-normal age [52].

Based on our own theoretical expertise and practice, combined with other relevant contributions identified through extensive literature analysis, we further identify **post-normal science** by **two categories of integrative sciences** (**Figure 4**).

First category consisting of two level 1 integrative sciences: a) **Systems Ecology** and nested (hierarchical) organization across space and time, focused on the structure (composition and relationships), ecological processes and functions, adaptive potential and resilience, and non-linear dynamics of the **Ecosphere' Natural environment** [41, 53–61], and gradually extended to the coupling relationships with social and economic compartments of the **human-dominated environment (social environment)**, which were identified as **Socio-Ecological Systems** (SES) [62–67] or as **Human and Nature Coupled Systems (HNCS)** [2, 68]; and b) **Sustainability science**, with its core of Ecological Economics, basically focused on the social and economic dimensions of the **semi-integrated 3D** pattern of

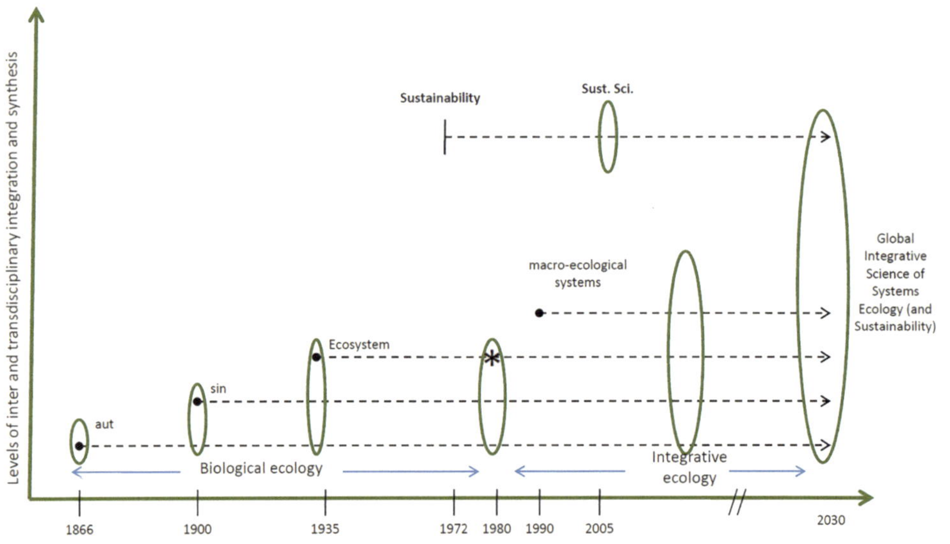

Figure 4.
Emergence and evolution of the integrative sciences: systems ecology and sustainability, and the global integrative science: ecology and sustainability of the nested macro-ecological systems of the Earth's life-supporting environment.

sustainable transition of the **nested socio-ecological systems (SES)** /or **human and nature coupled (HNC) systems** [3, 40, 69–71]. Therefore, we acknowledged that the evolution and synergy of these level 1 integrative sciences have provided consistent theoretical inputs for conceptualization, planning, implementation, monitoring, and governance of two decadal phases of the 3D sustainable transition launched at the UN Conference on Environment and Development, UNCED/Rio 1992/Agenda 21; Johannesburg 2002/MDGs; and third Rio2012 + and UN Sustainable Summit 2015/Agenda2030.

3.2 Level 2 integrative trans-disciplinary science

The second category refers to the emerging level 2 **integrative transdisciplinary science** of **Systems Ecology and Sustainability** of the **holistic** and **fully integrated** Ecological Complexes of the Earth's life-supporting natural and social environment (H-NSEC) [8, 72–74]. (**Figure 5**)

The sustainable transition of the nested Ecological complexes: Nature-society (ECNS) in the Lower Danube Sub-Catchment (LDSc), which expose high human-induced homogeneity and intensification of their biophysical and functional natural capital, demands new adaptive and transformative management and governance options consistent with a **holistic** and/or **fully integrated vision (attractor)** for **"Sustainable Future."**

In our opinion, the **core vision**, the **roadmap** consisting of a set of potential trajectories (pathways), and **operational infrastructure, methods, procedures,** and **tools** are the basic conditionalities (instrumental) for **long-term adaptive** and **transformative governance** of the **sustainable transition** across the spatio-temporal complexity of both the natural and social sides of Earth's life-supporting environment, beyond the current semi-integrated or coupled 3D (ecologic, economic, and social) vision and approach, acknowledged and practiced as 3D **sustainable transition** and, most often, 3D **sustainable development**.

Moreover, such an emerging theoretical and conceptual approach provides opportunities to identify and address the most stressing challenges and long-term solutions to many wicked problems that Humanity and Ecosphere with its nested H-NSEC$_s$ across space-time scales, will be facing during the sustainable transition in the new Anthropocene epoch.

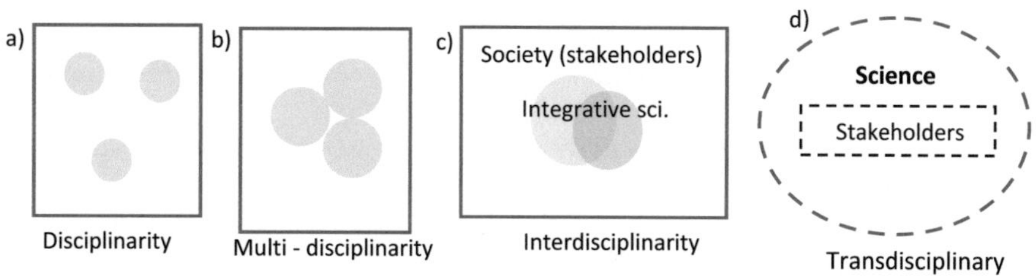

Figure 5.
Conceptual frames underpinning the evolution of integrative science: ecol and sust. A) Activity confined to one discipline; b) infers that two or more disciplines work in conjunction; c) fuses concepts, theories, and methods from different disciplines; d) goes beyond sciences, organizes processes of mutual learning between science and society, and innovates conventional patterns of knowledge, exchange, and provides platforms for sustainable governance.

4. Test case: Romanian lower Danube sub-catchment (Ro-LDSc)

Great rivers are important parts of the hydrological cycle in the Earth's natural environment, as well as the biophysical framework in which human societies develop. Both systems are closely related; socio-economic development depends upon resources and services supplied by the biodiversity and ecological diversity of natural capital at the scale of river catchments, and social and economic activities drive changes in the river system.

The river margins (ecotones or transition areas) between upland and water bodies have been identified and described as very sensitive and synthetic indicators of changes across river catchments [75–77].

Danube is the largest EU and second-largest European river, considered by the European Commission as "the single most important non-oceanic body of water in Europe and a future central axis for the EU." The Danube River flows from the Black Forest (Germany) along 2,857 km to the North-Western Black Sea, has more than 120 tributaries, and extends its catchment over 817,000 km^2. The water discharges of the river accounts for 60% of the total freshwater inflow to the Black Sea, or about 76% of the total inflow in the North-Western Black Sea. It is the world's most international river watershed, as it extends into the territories of 19 countries (ICPDR, 2000).

4.1 Test case description: Updated contextual framework

The Lower Danube stretch of 1,075 km long, serves as the border between Romania and Serbia, Bulgaria, the Republic of Moldova, and Ukraine (about 800 km), while the remaining portion (about 275 km) flows solely through Romanian territory. The Lower Danube Catchment (LDC) (**Figure 6**) extends over 218,600 km^2, of which 164,000 km^2 lies within Romanian territory (Ro LDC). The Ro LD Sub – catchment (**Figure 6**) covers an area of over 100,000 km^2 including 10,933 km^2 of wetlands (Ro LDW).

Table 1 shows that by the end of the extensive land reclamation campaign, most of the former wetland formations had been converted into human-controlled and energy-subsidized systems for crop and livestock production, intensive fish farming, or forest plantations in all subcomponents upstream of the coastal delta. The conversion of natural wetlands into arable land, forested areas, or fish farming systems has been much less intense in the coastal delta. Furthermore, after recent small-scale restoration projects and the abandonment of some former intensive fish farming systems, the composition and structural configuration of this area is determined mostly (more than 80%) by natural and seminatural or self-maintaining wetlands.

The empirical records from the early 20th century explain the wide recognition of the LDW system as one of the largest and most diverse wetlands in Europe. Despite minor hydro morphological transformations in the first half of the 20th century, by the end of the Second World War, the RoLDW system extended along the last 860 km stretch (currently downstream of the Iron Gate II man-made reservoir), preserving its structural configuration and functional capacity. The LDW system has been and remains one of the largest and most diverse wetland formations in Europe, covering a total surface area of almost eleven thousand square kilometers. This consisted of: i) the main river channel and arms, along with a wide range of shallow lakes connected

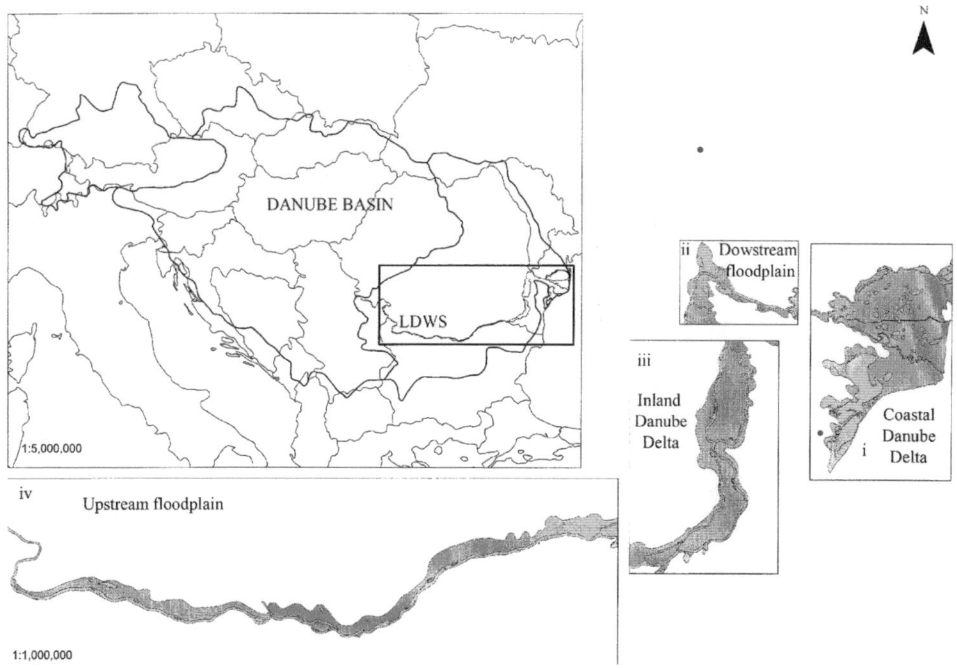

Figure 6.
Danube River catchment (Danube Basin) and four major wetland units of the Romania Lower Danube Wetlands (Ro-LDW).

Wetland units	River stretch (km)	Area (km^2)	Area of created polders (km^2)	% of change
Coastal Danube Delta (CDD)	0-71	4,467	1,041	21
Downstream floodplain	72-170	732	550	75
Inner Danube Delta (IDD)	171-370	2,926	2089	71.3
Upper floodplain	371-864	2,808	2037	72.5
	Total	10,933	5,747	52.4

Table 1.
Ro-LDW wetland units (see Figure 6), their surface area, and degree of conversion into agricultural land.

among themselves and with the river through a network of natural and man-made channels (approx. 25%); ii) wet meadows and marshes (30%); iii) grasslands and alluvial forests (27%) flooded 3–6 months annually; and iv) agricultural land (18%), which was usually flooded for a couple of weeks per year during major or extreme river pulses. Until the first decades of the 20th century, the traditional knowledge and practices of the local population allowed them to adapt the structure and metabolism of local and subregional socio-economic systems to the functioning regime and changes in the rates of resource and service flows, driven by the river's hydrological pulses and unexpected climatic and hydrological events [78].

However, significant geomorphological and ecological changes have been recorded along European rivers due to the implementation of policies and management programs designed in accordance with utilitarian principles and practices of the neo-liberal economy. In that regard, extensive and intensive human intervention in their structural configuration was designed and applied, such as – diverse hydrotechnical works aimed to straighten and deepen the river channel, to drain floodplain for agricultural land and physical constructions, to lock and dam the rivers to improve navigation, or to build large reservoirs for water resources and hydropower generation [79–82].

Although later than European rivers such as the Rhone and Rhine or the middle and upper Danube, similar utilitarian and reductionistic concepts and principles, complemented by sectoral management schemes and effective technologies, have been implemented in the LDWS [83, 84].

Moreover, the isolated and limited empirical data sets and information, produced by many small-scale, short-term, and focused studies, didn't help to avoid the extensive use of the utilitarian principles and sectoral approach in the fields of agriculture, civil engineering, and water management [4, 85–87].

4.2 Conceptual and analytical frameworks

In order to deliver **integrated, adaptive, science-based,** and **effective operational support**, which allows for understanding complexity across space and time, and to identify and obtain policy and decision-oriented solutions to wicked problems, we decided to focus on:

4.2.1 The instrumental role of the updated conceptual and operational frameworks for identification, planning, and assessment of the long-term dynamics within and between nested NSECs across the Romanian Lower Danube Sub-Catchment (Ro-LDSc).

In the case of the biophysical structure of the Romanian side of the LD Sub-Catchment, including the Ro-LDW and four regional **HNCS/SES**, the conceptual and analytical approaches evolved in parallel with theoretical integration and development in the fields of **System Ecology**, and **Sustainability** sciences, as briefly described above (**Figure 7**). The findings of long-term analysis (about 100 years) of such interlinkages allowed for the delineation of three major phases in the dynamics of conceptual, theoretical, and operational integration and synthesis (**Figure 8**). **The first phase** of conceptual and analytical approaches extended between 1900 and the late 1960s (**Figure 8**) and was basically dominated by poorly integrated theories of biological ecology, hydrology, hydrochemistry, and geomorphology. The research activities were mostly designed and implemented at small and medium (up to 10 and, respectively, 100 square kilometers) spatial scales and for short (months and years) and medium (5–10 years) terms, being focused on aspects regarding: i) population/species distribution and dynamics; ii) dynamics in species composition and structure of aquatic and terrestrial communities; and iii) river pulses, changes in water chemistry, and river morphology. The second phase of conceptual and analytical integration started at the beginning of the 1970s (**Figure 8**) as a response to the need to shift from sectoral, small-scale, and short-term approaches and reductionistic or discipline-based interpretation toward

DOI: http://dx.doi.org/10.5772/intechopen.1014638

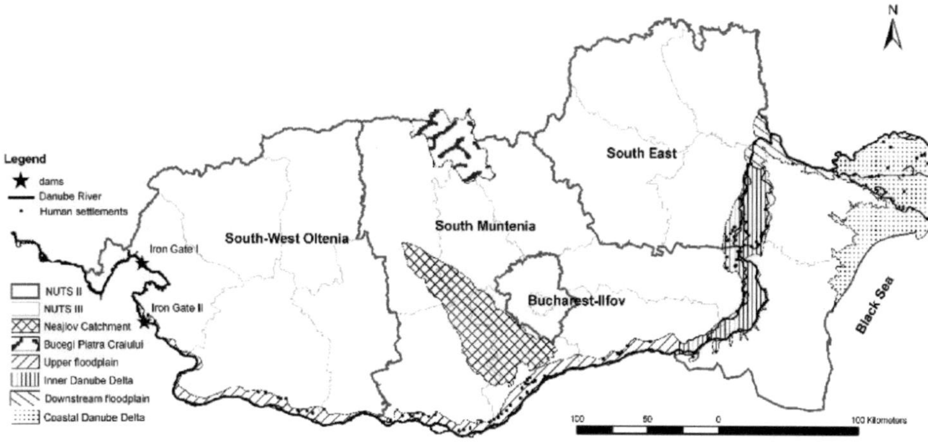

Figure 7.
Spatial distribution of the territorial administrative units (NUTS ii)/regional ECNs; LTSER sites and sub-regional and local ECNs in the Ro-LDSc).

ecosystem and landscape or hierarchical integrated approaches and holistic interpretation of the complex relationships among the physical, chemical, and biological components of the natural environment across Ro-LDW.

Such a conceptual and analytical shift has been enabled by the significant progress in theoretical integration and synthesis, consisting of the ecosystem and landscape as well as hierarchical organization theories. The theoretical foundation of new emerging sciences: systems ecology, and sustainability had a strong impact, since 1995, upon conceptual approaches and research program development and implementation (**Figure 8**).

4.2.1.1 Nested HNCS/SES in the Ro-LDSc and Ro-LDW

Since then, the research area has been extended on the nested local, subregional and regional Danubian human and HNCS/SES. In that regard four local HNCS/SES: i) Coastal Danube Delta (CDD) and Inland Danube (IDD) deltas representatives for the former and current structure of Ro-LDW; ii) River Neajlov catchment (RNJ), representative for the rural and agricultural area and; iii) Bucegi-Piatra Craiului, representative for mountainous and fast economic development area of Ro-LDC (**Figure** 7) have been established as the core of national network of Long Term Socio-Ecological Research (Ro-LTSER). As well as, **test case** in more than 15 EU Research Projects among which more important for this paper are [89–92].

Figure 8 shows that for almost a century, the conceptual model has been designed and improved, aiming to provide frameworks for identifying and understanding the structural and functional complexity of Ro-LDW at population/species, community, ecosystem, and landscape scales. In terms of Bio and Eco diversity, the conceptual framework allowed for the clarification and definition of the biophysical structure and physical economy of natural capital as the natural footprint for the socio-economic system. In terms of capital, the conceptual models, which were extensively used in the first and second phases of long-term field research, have been established to integrate the full range of key components from the biophysical structure of

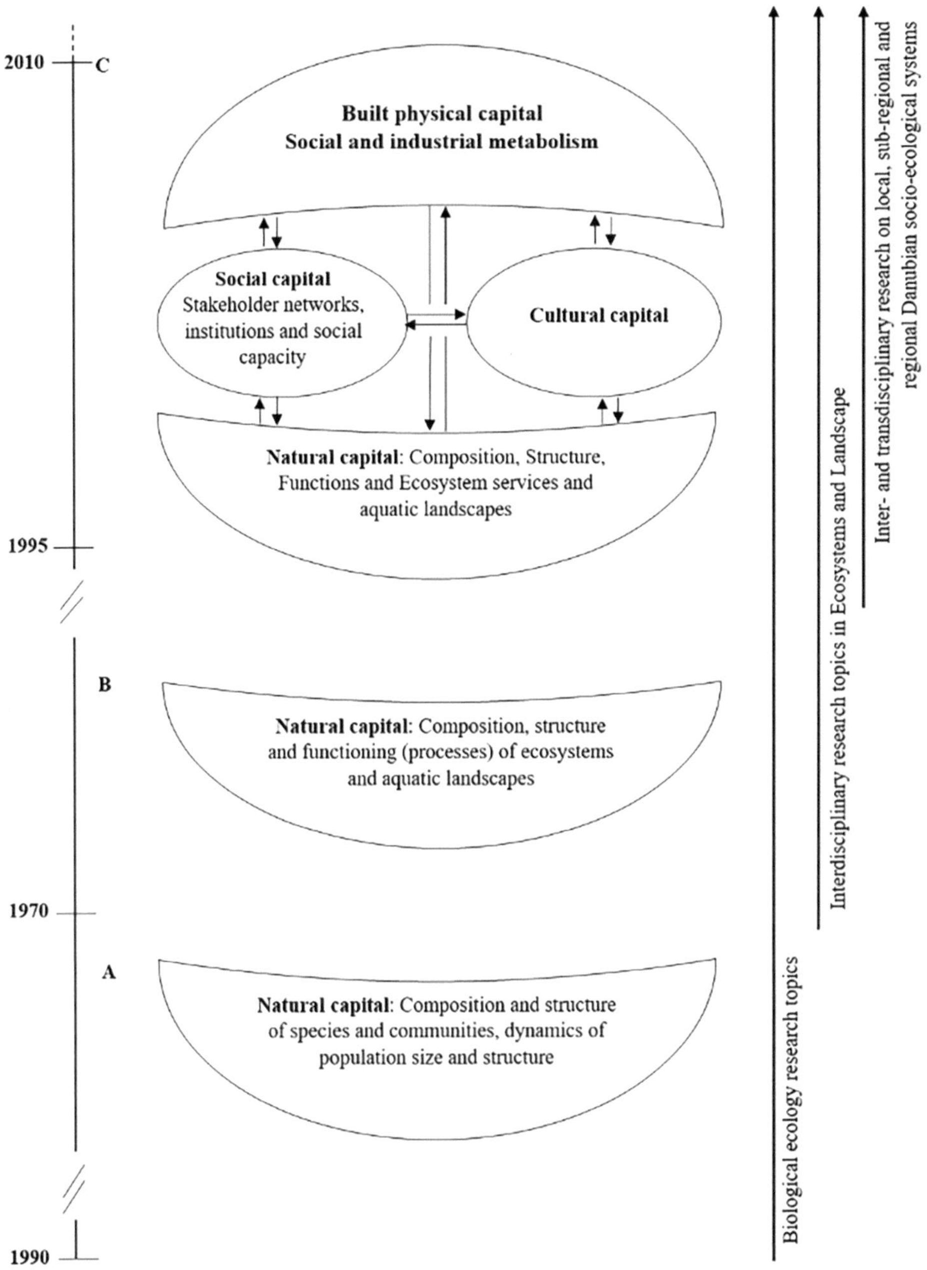

Figure 8.
Successive phases in the development of operational framework in the long-term research of the Romanian Lower Danube sub-catchment (after [88]). A. Small-scale and short-term bio-ecological research; B. Long-term research on ecosystems and aquatic landscape in the coastal area and in the inland Danube Delta; C. Designing and starting to carry out long-term and large-scale socio-ecological research.

natural capital/biodiversity. Thus, a wide range of data sets and information have been produced and are currently used for integration and synthesis, aiming to understand and evaluate the integrated ecological impact of policy, economic, and social stressors across the Danube River watershed, specifically those from the Low Danube Catchment on the Ro-LDW (including the Danube Delta Biosphere Reserve and Razim-Sinoe Lagoons) during (1950–1989) accelerated socio-economic development [83, 86, 93–110].

4.2.1.2 Generic conceptual framework for identification Danubian HNCS/SES

In the last phase of conceptual integration and development, we identified the Danubian HNCS/SES (socio-ecological systems) through a conceptual framework, which integrates three other compartments that describe, in terms of the capital concept, the **biophysical infrastructure** (Built and Social capital) and the **symbolic realm** (cultural capital) of the socio-economic subsystem, as well as the interactions and interdependencies among them. This was further improved and converted into a more detailed conceptual framework (**Figure 9**) which better reflects the structural and functional complexity within and between nested 3D – HNCS/SES across the entire catchment of the Danube River (DRC) and allowed for significant operational development.

First rethinking and developing the research program and infrastructure in order to integrate new research areas and topics, focused on the dynamics of complex interactions between the components of socio-economic systems and that of Natural Capital (e.g., biodiversity and ecosystem services identification, assessment, and valuation; people's perceptions, attitudes, and behavior related to biodiversity/natural capital conservation, sustainable use, and restoration; socio-economic drivers and pressures), as well as for diversifying and improving the research activities across space at all levels of organization (population/species, communities, ecosystems, and landscapes) of the biophysical structure of Natural Capital, aiming to extend the knowledge about: i) the contribution of populations/species to ecosystem and landscape processes, functions, and services; ii) structural thresholds and tipping points in the dynamics of ecosystems and landscapes; iii) key natural and socio-economic drivers and pressures, including climate changes, which drive their long-term structural and functional dynamics; and iv) the relationships between biodiversity and ecosystem/landscape services and resilience.

Second designed and test: the operational platform support (**Figure 10**) for adaptive and transformative governance of 3D-Sustainable transition (3D-ST) of the nested $HNEC_s$/SES across Ro-LDSc and RoLDW, according to theoretical and operational achievements of Level 1 integrative sciences [111].

Third also allowed for designing and adopting the comprehensive framework (**Figure 11**) which reflects the theoretical development of **post-normal integrative trans disciplinary and transformative science – System Ecology and Sustainability**, as well as the architecture of the **integrated Science-Society-Policy Platform (SSPP)** for adaptive and transformative governance of the **Sustainable transition (ST) of nested h-NSEC$_s$.**

In conjunction with these developments, we carried out long-term data sets analysis, integration, and synthesis, aiming to develop the first generation of inter and transdisciplinary knowledge concerning past inter related and transformative changes, and their trade-offs recorded between 1950 and 2020, with a focus on Ro-LDW as an impact integrator of many long-distance and cumulated drivers and pressures.

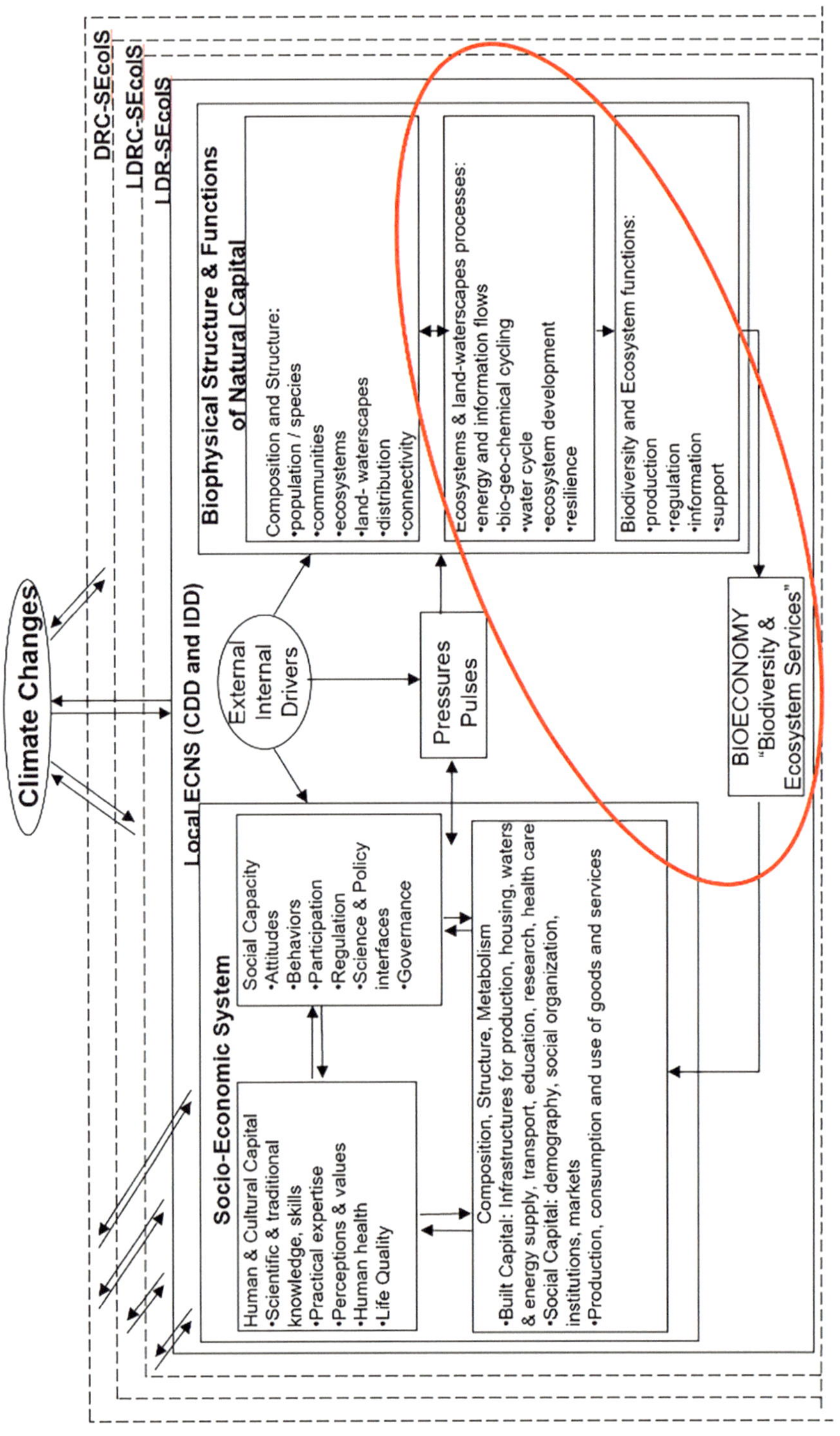

Figure 9.
Conceptual framework on the organization and relationships between nested ECNS from the Danube River catchment (DRC).

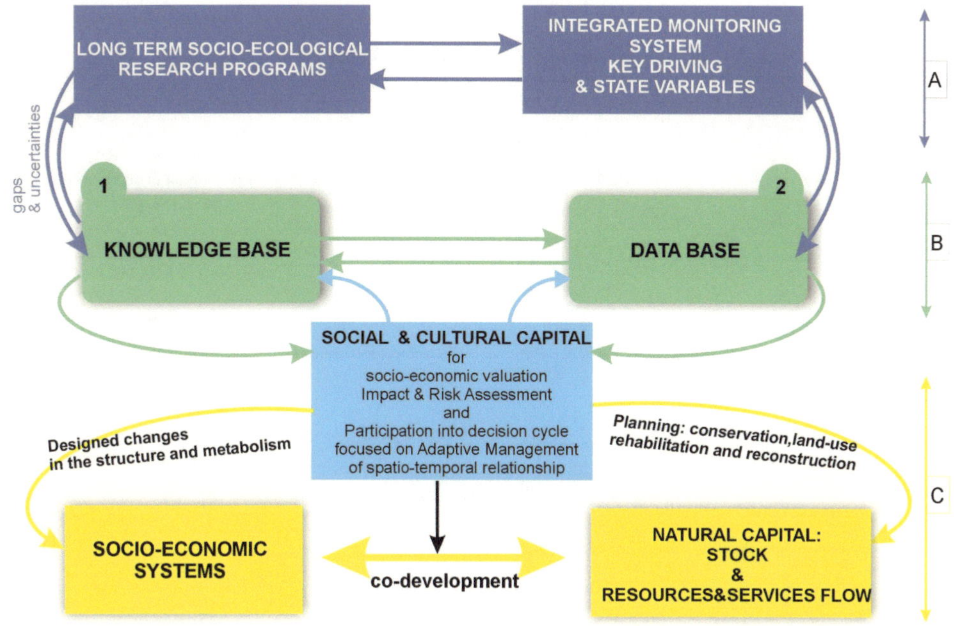

Figure 10.
The operational platform support for adaptive and transformative governance of 3D – sustainable transition (ST) of nested HNEC/SESs: (a) research and monitoring infrastructure; (b) information system (core of decision system); and (c) adaptive management for co-development (sustainability).

4.2.2 Social and economic transition

One of the main targets of the EU-VOLANTE research project was addressing the need to understand the historical transformative transition of European societies and their natural environment.

Our narrative regarding the dynamics of major drivers – technological, institutional, and economic – of long-term land use changes (1860–2010) was an attempt to provide an integrated overview of the process in the case of Romania.

We identified six land management regimes coupled with three socio-economic transitions: i) **first capitalist transition** (1863/1945) from a rural society and agrarian economy of feudal type toward democratic societies and a capitalist economy-private ownership, capital accumulation, and free market; ii) **socialist – communist transition** (1947/1989) from a very young and vulnerable capitalist society and economy toward a socialist–communist society, shaped according to the imported communist ideology and a socialist economy designed, built, and managed in accordance with established sets of norms and targets – state-owned land and physical capital; central planning for major institutional and economic objectives, particularly those concerning industrial, agricultural, and urban sectors; command-and-control management regime applied in the implementation of the approved programs and plans; iii) **Second Capitalist transition** (1990/2007 and after EU integration).

In line with the aims and objectives of this chapter, we carried out a brief analysis and interpretation of major social and economic transformative changes

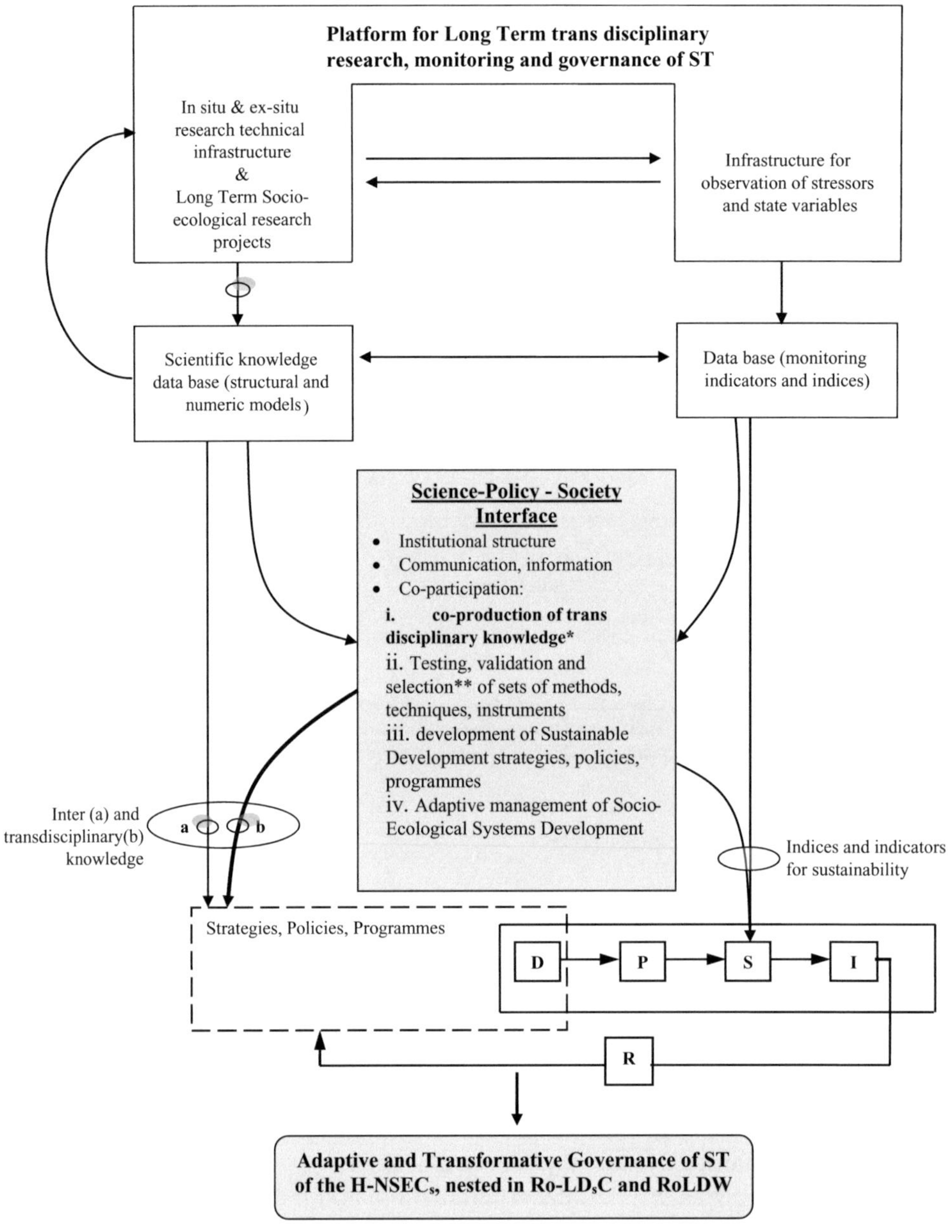

Figure 11.
Platform for long-term inter and trans-disciplinary research and monitoring of nested h-NSECs – Ro-Lower Danube sub-catchment, support for adaptive and transformative governance of ST.

and the related drivers (institutional, technological, and economic) for the land management regime of the Second (socialist-communist) and Third (capitalist and sustainable) transitions in the Ro-LDSc, including Ro-LDW, and their past and future cumulated trade-offs reflected by selected, most sensitive indicators and scenarios.

4.2.2.1 Social-communist transition and land management regime: Ro-LDSc and Ro-LDW

The land management regime was defined by: cutting off land property rights and expropriating the former medium-sized (30–100 ha) private land – properties and the land owned by cultural and spiritual institutions; extensive land reclamation; concentration of former family exploitation, and establishing large and intensive state-owned farms (average 4,800 ha), which extended over 20% from 7,2 million hectares of agricultural land, and collective farms (average 2100 ha) exploitation (74% of agricultural land) and a set of specific characteristics such as – fast process of urbanization and industrialization; market-oriented production; state and collective farms established (after 1975) specialized units for crop and livestock production; and, 10 years later, some of the largest state farms established independent and specialized livestock production. This transition was driven by effective technological drivers – access to modern equipment and science-based technologies; selection of resistant and productive breeds; inorganic fertilization, irrigation, and chemical pest control; Institutional drivers – state and collective rights on 90% of agricultural land, specialized human resources, central planning, and control. Economic drivers were characterized by a dynamic internal market related to the increasing urban population, which grew from a formal level of 20% to up to 60% in 1980's and the increasing demand in external markets.

The large state farms were established by expropriation (1945 and 1949) of former landowners or by extensive land reclamation of about 7,000 km^2 of which 5,732 km^2 were in the Ro-LDW [112].

The 12 years of aggressive implementation of agricultural policies, which allowed for the expansion and aggregation of arable land into large state and collective farms, have been followed by 28 years of large investments in physical capital – irrigation and drainage infrastructure for about 2.8 million hectares of arable land, centers of agricultural equipment for a wide range of agrotechnical works intended for intensive crop production (e.g., from 6,000 tractors in 1950 to about hundred thousand in 1980s), and increasing the inputs of inorganic fertilizers up to 700 Ktons per year.

By the end of 1980s, the rural areas of Ro-LDSc and Ro-LDW consisted of homogeneous agricultural landscapes, with very limited and fragmented semi-natural habitats, intensive metabolism, and a high dependency on commercial energy and materials (e.g., diesel, fertilizers, pesticides).

Such structural configuration and intensive functional regime of engineered agricultural landscapes were accompanied by a wide range of cumulated externalities (e.g., soil erosion and exploitation reflected by 15–25 t of soil annual loss per hectare), diffuse nutrient emissions estimated at 50% of the total contribution of Ro-LDSc to nutrient load discharge in the Danube Delta and North Western Black Sea.

Our brief analysis of the socialist-communist transition in the Ro-LDSc indicates land reclamation and diffuse nutrient emission as major drivers in land use planning and cumulated and long-distance externalities (trade-offs).

For better understanding and recognition of the importance of nested and holistic analysis at large space and time scales, we have selected sensitive indicators of trends, thresholds, and tipping points of long-term changes in the biophysical structure and functioning capacity of Ro-LDW.

4.2.2.2 Synergy and trade-offs

Figure 12 shows the synergy between changes in the structural configuration of Ro-LDW due to extensive and intensive land use changes (see **Table 1**), and **changes** in the **aquatic trophic condition** driven by the **increasing diffuse** and **point nutrient emissions** from primary and secondary sectors of the emerging industrial economies in the Ro-LDC, reflected by the **dynamics of the N:P ratio**.

The high level of nutrient load in the Danube, recorded at the end of the 1950s and the beginning of the 1960s [96], seems to relate to the intensification of crop production, social and industrial metabolism, and the extensive conversion of wetlands upstream of LDC.

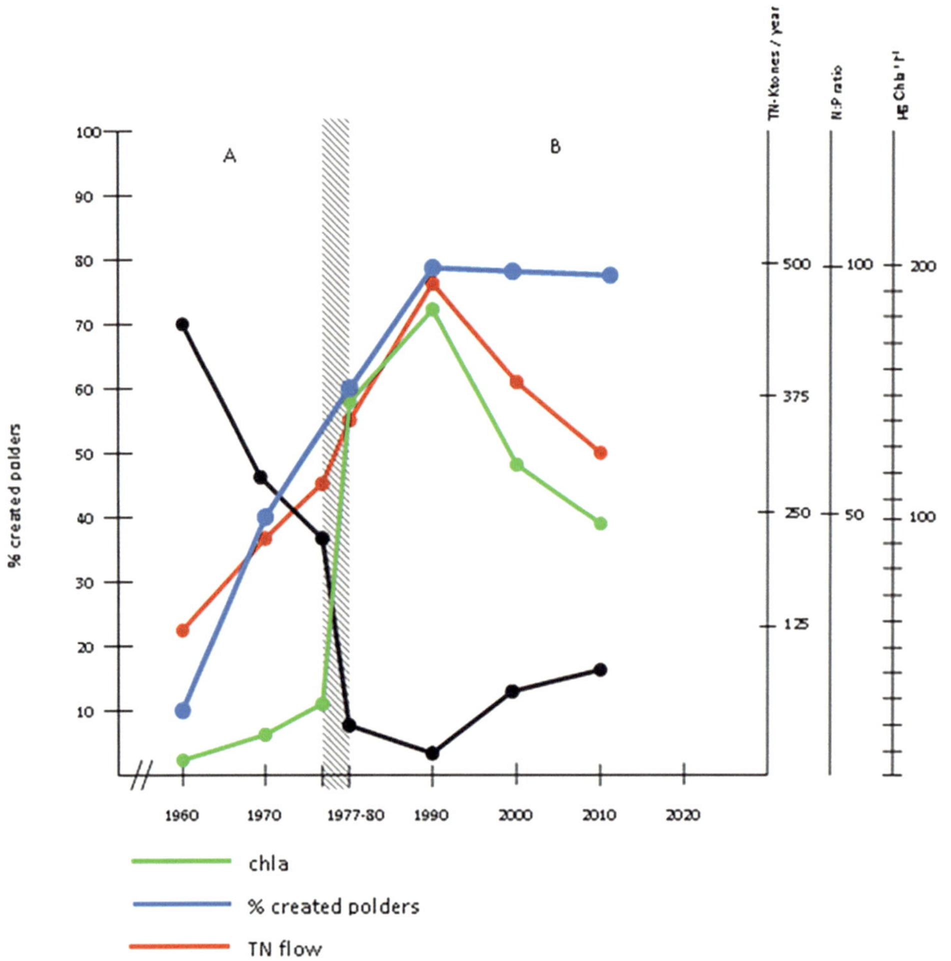

Figure 12.
Synergy and trade-offs between changes in the structural configuration and aquatic trophic state of Ro-LDW.

However, the maintenance of almost full structural integrity of the Ro-LDW until 1956 made it resilient to the high density of nutrient inflow from the mid and upper DRC.

Moreover, the Ro-LDW preserved its high buffering capacity for nutrient inflow for nine years, between 1956 and 1965, despite large structural changes in the 3rd and 4th subunits (see **Figure 6**). The result consisted of 3,600 km^2 of agricultural polders, equivalent to 33% of the total surface of Ro-LDW.

The dynamics of N:P ratio and Chl.a reflect the capacity of the Ro-LDW, even after significant structural changes and persistent high nutrient inflow, to preserve meso-trophic conditions and related functional regime, but, since the mid-1960s, the system became vulnerable either to more wetland conversion or an increase in nutrient inflow.

The fluctuations of N:P ratio and Chl.a during 1965–1980 indicate that the resilience of the system has been exceeded, as well as a relatively smooth transition toward eutrophic conditions.

The narrow range (34–38%) of increase in polders presence in the Ro-LDW has been identified as a structural threshold, which allowed successive shifts toward eutrophic and hypertrophic functioning regimes of Delta's aquatic ecosystems.

4.2.2.3 Interactions and trends of key land-waterscape and lentic ecosystem services

The shift in the structural configuration and functional regime of the Romanian Lower Danube Wetlands (Ro-LDW) generated significant changes in their support regulation, provisioning, and information functions, as well as in the ecosystem services upon which local and subregional populations and economies depend.

Altough a wide range of services have been identified for both Ro-LDW and its lentic ecosystems, our analysis focuses on six key services that are most relevant for regional socio-economic systems and for which reliable qualitative and quantitative data are available. These include: one support service (Net Primary Production - NPP); three regulation services (river pulse regulation, nutrient flow regulation and water purification); one provisioning service (fish yield); and one information service (recreation and ecotourism).

We assume that nutrient flow regulation provided by Ro-LDW acted as a primary driver of structural and functional changes, with cascading effects on the shallow lakes of the Danube Delta. As illustrated in **Figure 13**, all selected services - except NPP - declined to varying degrees relative to reference conditions. The extensive conversion of floodplain areas (52.4%) into energy - and nutrient-subsidized systems designed for intensive exergy accumulation (crops, timber, and fish biomass) resulted in a reduction of more than 75% in water storage capacity and river pulse regulation.

Figure 13 presents the intaeraction network among key Ro-LDW services. A strong negative interaction is evident between river pulse regulation and nutrient flow regulation. The reduction of floodable surface areas not only weakened the capacity to buffer hydrological pulses but also reduced nutrient retention capacity, thereby facilitating nutrient enrichment in the Coastal Danube Delta and the northwestern Black Sea. Both regulation services show strong positive interactions with the main support function - exergy accumulation expressed as NPP - which sustains aquatic food-web networks.

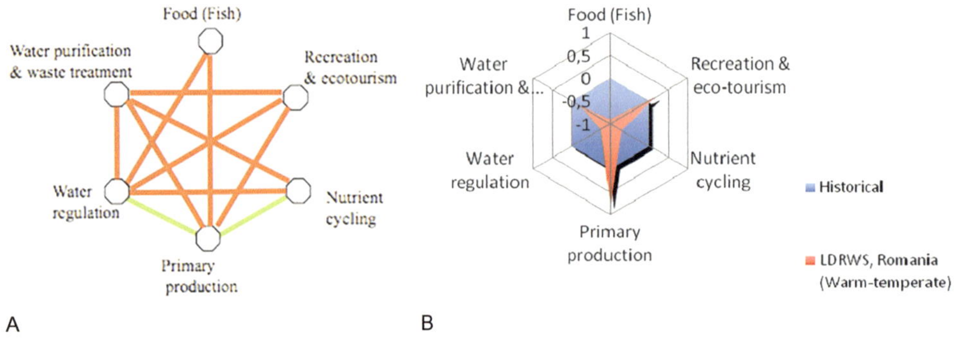

Figure 13.
Interactions (a) and trends (b) of key ecosystem services provided by the Ro-LDW.

At the Ro-LDW scale, pronounced negative interactions emerged between river pulse and nutrient flow regulation services, on one side, and water purification and waste treatment services, on the other. The analysis suggests that the degradation of hydrological and nutrient regulation actyed as a systemic driver, producing negative impacts on water purification, provisioning (fish yield), and information services (recreation and ecotourism), while simultaneously stimulating increases NPP within aquatic ecosystems. **However, elevated NPP levels intensified structural imbalances in food webs and exacerbated water -quality -related pressures, with adverse consequences for bathing water, drinking water supply, fisheries productivity, and ecotourism.**

4.2.2.4 The second capitalist transition at the roots of the reverse land tenancy configuration (1990–2010)

By the end of 1989, the Romanian socio-economic system, restructured and managed between 1948 and 1989 according to the communist principles and practices, collapsed and entered a new transition process from the communist-socialist structural configuration and governing regime toward a new attractor, defined by capitalist concepts of private property and free market economy, as well as the principles and norms of advanced Western democracy. For that, the former Constitution and most restrictive laws and norms were abolished, and a new Constitution (1990) and the first set of laws for land reforms were adopted in 1990/1991, in order to allow for designing the transition framework and attractor (**Table 2**).

Characteristics:

- Land property restitution
- Fragmentation of the large-scale exploitations
- Arable land abandonment
- Dominance of subsistence and semi-subsistence exploitations

Technical	• **Mechanization: Prohibited for most subsistence exploitations** • **Deterioration of the irrigation system** • **Less mineral fertilizers** • **Dominance of non-intensive agricultural production** • **Loss or deterioration of the infrastructure for intensive livestock production**
Institutional	• Co-existence of subsistence and semi-subsistence exploitations, commercial societies (CS), and first set of agricultural societies (AS) • Land rental and transaction market
Economic	• Lack of subsidies • Subsistence production • Expensive credits • Disturbed internal market and vulnerable to imported agricultural products

Table 2.
Transition from the socialist to capitalist socio-economic system and specific "land management regimes": Land management regimes: reverse land tenancy and fragmentation of state and collective exploitations.

Drivers:

The reverse transition of the Romanian socio-economic system toward capitalism and EU-accession has been accompanied by three successive transitions in the land management regime: i) Main policy objectives during first land management regime – "reverse land tenancy" have focused on developing the specific regulatory framework and instruments which allowed: a) distribution of small (0.5 ha) parcels of land to all members of agricultural cooperatives who did not own land before collectivization; b) restitution of land property rights to all former owners or their heirs; and c) emerging a functional land rental and transaction market [112, 113].

The first phase (1990–2000) of the second capitalist transition, reflected in the agricultural sector, could be characterized by: i) the shift from large-scale farms to small subsistence farms; ii) a sharp decrease in the level of inputs (e.g., quantities of fertilizers employed were reduced by more than 70%, iii) irrigation and drainage systems built during the communist transition were either destroyed or deteriorated (e.g., only about 15% of the area previously equipped was or is currently irrigated); iv) a high rate of inflation, unstructured and low functional efficiency of the market (including the rental market); v) lack of subsidies to compensate for large disparities between the cost of production and benefits; vi) very low access to machinery services for the small individual farms [5].

Under these circumstances, the agricultural production displayed a serious decline and a high level of uncertainties (e.g., 60% reduction in livestock; variable cereal production within the range of 8–14.8 million tons).

The second land management regime, or the preparatory process for joining the EU (2000–2006), has been driven by two new policy objectives: i) transposition of EU-CAP policy objectives, mechanisms, and standards, and the related directives (e.g., WFD, HD, BD, Nitrates Directive) into domestic policy for land and rural reform and the regulatory framework for agro-environment, and ii) creating specific organizations and institutions and building capacity aimed at the implementation of Common Agricultural Policy (CAP) objectives and financing schemes in the post-accession phase. – (Table 3)

Technical:	• **Mechanization** • **Mineral fertilization** • **Irrigation**
Institutional:	• Agri-environmental schemes • Structural Operation Programme/Axis 4 • Technical assistance and training
Economic:	• Area – based subsidies • High cost of land transactions and credits • Market regulation and improved conditions for semi-subsistence farmers to enter and operate in the market

Table 3.
Restructuring and improving the viability of agricultural exploitations in the pre and post-EU accession process: Two pillars – increasing environmental awareness.

Characteristics

- Transposition of EU-CAP, Habitat Directive (HD), Species Conservation Directive, Nitrogen Directive, and Water Framework Directive;
- Development, testing, and consolidation of the institutional and regulatory frameworks and capacity;
- Changes in the composition and structure of the agricultural landscape by improving the effectiveness of land rental and transaction markets;
- Implementation of the environment and biodiversity standards in land management practices (in particular, Natura 2000 sites);
- Stimulating the shift from subsistence toward semi-subsistence and medium-sized (20–25 ha) market-oriented exploitations;
- Increase the number of young and professional farmers;
- Polarization of land uses: intensive and monofunctional vs. extensive and multifunctional farming.

Drivers:

As an EU member state since 2007, the strategic policy, social and economic objectives defining the attractor of the Second Capitalist transition and management regime have been adapted and improved to bring clarification, coherence, and efficiency to this complex, challenging, and long-term process, and by doing so, to transform the vision (attractor), road map, and social, institutional, technological, and economic tools and procedures into a semi-integrated (3D) and holistic and nested conceptual (see Section 2) and operational framework for sustainable transition and adaptive and transformative governance (management regime).

In that regard, the second National Agriculture and Rural Development Program (2007–2013) and the first National Strategy and Action Plan for Sustainable Development (NSSD) 2008 have focused on the transformation of rural landscapes, dominated by subsistence farms and deteriorated built infrastructure, into ecological, economic, and socially viable rural socio-ecological systems, as well as the Romanian socio-economic system and biophysical structure of Natural Capital into nested sustainable socio-ecological systems.

Therefore, we assume that the nested NSEC^{s}: Ro-LDSc and Ro-LDW undertake long-term (horizon time 2050) sustainable transition and governance regime in order to reduce the social, economic, and ecological costs, and increase the benefits and well-being of the on going "second capitalist" transition. Our assumption takes into consideration the many administrative, political, social, and economic wicked issues, power relations, and conflicts, but also valuable and convincing outcomes in the long-term Sustainable Transition (23 years) of DDBR (Danube Delta Biosphere Reserve) as one of the pilot local and subregional human and nature-coupled ecological complexes (HNEC) for testing and developing the knowledge, expertise, and practice of the 3D-conceptual framework of Sustainability.

We learned that long-term sustainable transition and adaptive and transformative governance depend on the identification of risks and uncertainties, and the capacity to find viable solutions, long alternative pathways (road maps), as well as the availability of diverse operational tools, among which scenarios:

In our test case, the scenario exercise was carried out based on: i) the concept of integrated and hierarchical organization across space and time scales of the natural and social environment into nested social-ecological systems; ii) the conceptual framework and methods applied by the EU-Openness scenario team; iii) the retrospective analysis and synthesis of empirical data series and information, which allow understanding the changing pathway, current key state attributes, and challenges; and iv) participation of a wide range of stakeholders.

5. Well-being scenario (WBS): Nested Ro-LDW and Ro-LDSc

5.1 Trigger of changes

Restoration, rehabilitation, and development of the physical infrastructures for intensive industrial and agricultural production activities; transport; and urban and rural settlements to improve living conditions; raising the basic economic indicator (GDP) up to or even beyond the 2/3^{rds} of that of Western EU countries (**Figure 14**).

1. Well-being (WB); 2. United-we-stand (UWS); Sustainable transition (ST)

5.1.1 Political, societal, and economic changes:

There are many expectations that, under internal (i.e., increasing social capacity and societal demands) and external (EU policies and regulations, which are expected for the implementation of the EU Well-being scenario) pressures, the ongoing slow,

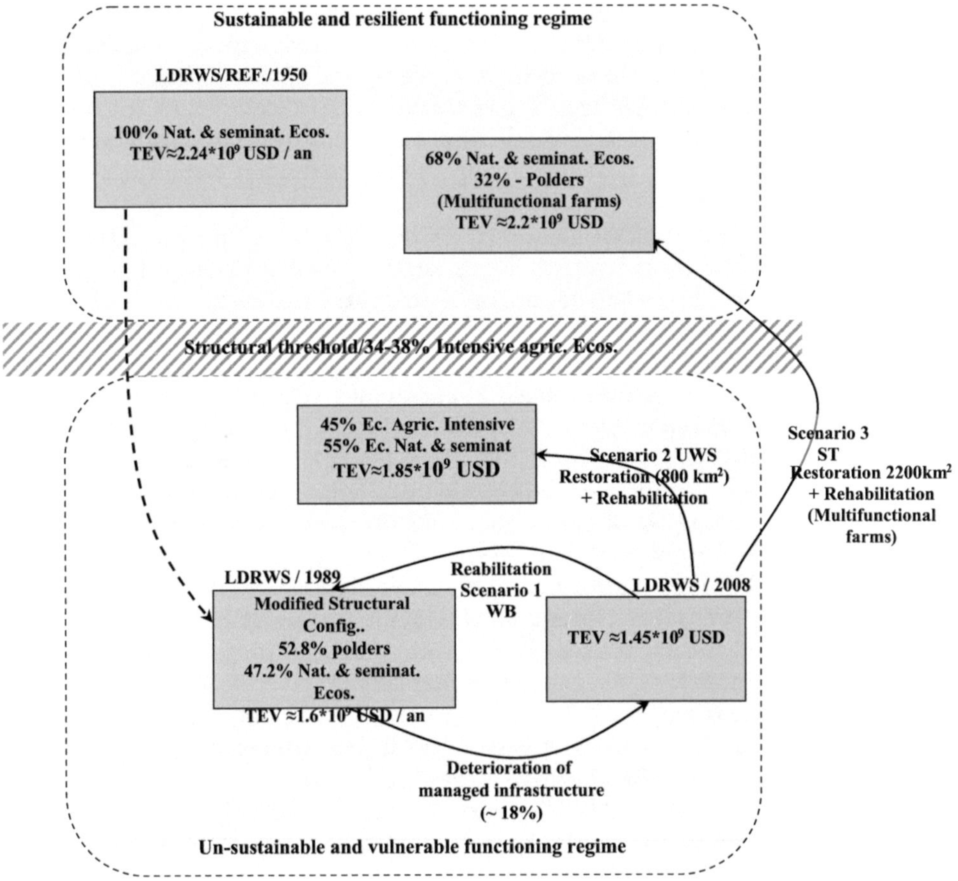

Figure 14.
Alternative scenarios for two governance regime of the second capitalist transition of nested NSECs in the Ro-LDSc: Ro-LDW as trade-offs integrator and buffer of land use and diffuse nutrient emission drivers.

confused, and highly corruption-dominated process of "economic reform" and changes in organizational and institutional arrangements and capacity will be improved in order to enable decentralization and "effective" governance at all policy and decision levels (national, regional, and local). The sectoral approach we are expecting is likely to generate many power and management conflict issues.

Main drivers of economic growth: i) restoration and development of physical infrastructure for intensive agriculture and industrial production capacities; ii) consumerism (high domestic and mainly external demand for food, chemical products, biomass, and bio energy); and iii) raising the quality and efficiency of the domestic economy and living standards to levels comparable with those of EU countries, along with full EU integration by 2050.

Restoration of former and/or development processing industrial capacities (i.e., ferrous and non-ferrous metallurgy; petro-chemistry) in the lower Danubian cities will require between 40 and 70% import of raw materials and fossil fuels, while almost 2/3 of the production of these industries will be exported.

The profit in this sector will be highly dependent on low labor costs and investment in pollution abatement, compared with Western EU countries. An economic growth of

4–5% annually is expected to be achieved mainly through the contribution of primary economic sectors: i) extraction and processing industries and ii) intensive, highly material and energy-subsidized agriculture, practiced in large (usually > 1000 hectares) exploitations; aquaculture (most reservoirs distributed in the Lower Danube floodplain – Ro-LDW), biomass production in intensive forest plantations, as well as industrial livestock farms (> 50 thousand capita).

Contributions are also expected from the food industry, tourism, renewable energy, transport, and trading.

In any strategic and policy documents, the importance and need for appropriate public investments in the education and health care sectors will be outlined, but they are expected to remain far behind the western EU level, at least until 2030. Similar gaps are also expected in the average salaries and living standards.

Under these circumstances, the emigration of highly educated people (i.e., medical doctors, engineers, IT experts, and scientists) will remain a critical challenge that must be carefully managed until 2030, and from then on, a more significant part of the benefits of economic growth toward creating comparable living conditions to those in developed EU and North American countries. Additionally, efforts should be made to establish a balanced movement (probably after 2040) of the labor force among EU countries and regions.

5.1.2 Urban, rural, and gray infrastructure: Restoration/rehabilitation and development

1. The infrastructures and production capacities consist of: 1.1) Industrial units: ferrous and non-ferrous metallurgy (>12 million tonnes per year production capacity); chemical and Petro-chemical products, including fertilizers, pesticides, and pulp and paper; shipyards – for maritime and fluvial shipbuilding; food and wood industries; 1.2) Agricultural infrastructures: rehabilitation of the flood protection system for more than 5,000 km^2 of polders established between 1960 and 1989 in the Ro-LDW; irrigation and drainage systems inside polders and outside lowlands; restoration of large and intensive agricultural exploitations; restoration and development of infrastructures for intensive fishery and large animal husbandries; 1.3) Development of infrastructure for renewable energy generation: windmills, photovoltaic fields, and biofuels;

2. Rehabilitation and development of the transport infrastructure – waterway and harbor facilities as part of the EU waterway linking the North and Black Seas, railways, and roads;

3. Rehabilitation and development of physical infrastructures in the urban and rural settlements – water supply and sewage systems; waste water treatment plants; solid waste collection and disposal or recycling; urban transport; health care and education; recreation.

5.1.3 Land use and environmental conservation

After more than 35 years of a second capitalist transition from state-owned and centralized governance toward a private-based and free-market economy, the sectoral policies and management plans for economic development are shaped by the same principles of utilitarian neo-classical economics applied during socialist governance in Romania and elsewhere. Accordingly, the floodplains are considered “wastelands” that have to be converted into productive agricultural land.

Large (hundreds and thousands of hectares) mono-functional, crop-specialized, and intensive private agricultural exploitations will replace the current commercial societies and hundreds of thousands of subsistence (2–5 ha) and semi-subsistence (6–10 ha) farms, established during the economic transition in the Ro-LDSc and Ro-LDW.

Diffuse nutrient emission from agricultural areas will once again become one of the major stressors (≥50%) responsible for high nutrient loads and discharges (≥400 ktonnes of Nitrogen and 30 ktonnes of Phosphorus) from the Danube River into the North – Western Black Sea, while maintaining the extensive changes in structural architecture/land use in the Ro-LDW (78%/4,900 km^2 of polders up-stream CDD) and thus causing a significant reduction in regulation, information, support, and production functions (e.g., habitats for fish species spawning and feeding; extensive fishery).

Although the management of the biophysical structure and functioning capacity of Ro-LDW and the Lower Danubian Plain (part of Ro-LDSc) should be designed and implemented in accordance with the provisions of many strong conventions and directives (i.e., Black Sea, Danube River Conventions, Ramsar, WH Conventions, and EU Directives – WFD, Habitat, and Birds Directives), it is more likely that the effective measures for achieving the goals will be systematically postponed, “forgotten,” and underfunded, at least during the restoration and rehabilitation of economic infrastructures.

6. United we stand (UWS) scenario

Restoration and/or rehabilitation and development of the physical infrastructure for intensive agriculture and industrial production activities, and fast economic growth will be like in the scenario – wealth being the basic target of the UWS scenario. In addition, specific actions and investments will be targeted toward:

i) Restoration and conservation of the biophysical structure of the Ro-LDW to supply key intermediary and final ES^s;

ii) Development of social capital and social well-being at levels comparable with those of other developed EU regions and countries.

The UWS scenario takes into consideration a very important measure, envisaging partial restoration of the Ro-LDW. In that regard, it is proposed that wetlands restoration to be extended over 800 km^2 from IDD and upper floodplain units. The rehabilitation of polders and associated built infrastructure is maintained as a core measure. Like in the first scenario, the assumption concerning the control of nutrient inflow is accepted. It can be noticed that, in this case, there is an attempt to address both pressures, which have been proved to be responsible for the structural and functional transition of Ro-LDW.

Our estimations suggest that the structural configuration remains in the same basin of attraction, preserving high vulnerability to external stressors and a significant negative balance of physical economy and monetary flow.

7. Sustainable transition (ST) scenario

The sustainable transition scenario proposes, as a major target, the reversible transition in both the structural configuration and functional regime of Ro-LDW. According to our estimation, this can be achieved by: i) restoration of 2,200 km^2 of wetlands, primarily upstream of CDD, which can ensure the shift of structural configuration back to the reference basin of absorption; ii) rehabilitation of soil quality and built infrastructure in the remaining polders (≈3,547 km^2); iii) promotion of multifunctional farming systems in the remaining polders; and iv) halving the nutrient inflow contribution of Ro-LDSc recorded in the late 1980s (27% or ≈115 ktonnes of TN per year and 23% or ≈10 ktonnes of TP per year) by implementing efficient measures addressing the socio-economic metabolism of lower Danubian socio-ecological systems. We consider the ability of policy and decision-makers to create and maintain a balance among all types of proposed measures as the most critical factor for achieving the goal of transition. In fact, we propose, for public debate and decision-making, two unsustainable and one sustainable scenario. This provides a broad framework and knowledge support to a wide range of stake holders for critical assessment of the short- and long-term costs and benefits of different options, which further creates opportunities for sustainable management of Ro-LDW and sustainable development of nested socio-ecological systems from Ro-LDSc.

8. Conclusions

The paper addresses one of the most challenging thematic areas, dealing with land and open water use planning, and adaptive and transformative governance of the sustainable transition in the new human-dominated geological époque – Anthropocene.

The transition is a global, complex, non-linear process, and for its understanding and governance across space and time scales, we propose a holistic approach based on retrospective and prospective contextualization, theoretical and conceptual evolution, and operationalization. In that regard, we consider the contextual, conceptual, and operational frameworks as fundamental for wicked issues identification, analysis, and defining alternative and transformative strategies, policies, and actions. Being aware of the many constraints responsible for the proper understanding, designing, and governance of sustainable transition, we organize the chapter into three major parts focused on: a) Historical contextualization of the origin, distribution, and dynamics of the global biosphere and ecosphere as a Life Supporting environment (**Figure 1**) which can help in better understanding the structure of the Earth System, the position and structure of the ecosphere and climatic system, and the major contribution of the biosphere to the relative stability of the ecosphere in the last geological epoch – Holocene, and thus for the evolution of human societies. It also helps clarify many concepts, among which – environment, nature, social space, natural/ecological space, the biosphere and ecosphere boundaries, as well as aiding in the identification of the global ecological crisis as a driver and starting point for long-term sustainability transition;

b) Theoretical and conceptual background. Here, we provide a brief but consistent synthesis (**Figure 4**, **Figure 5**) on the fundamental theoretical and conceptual development of two categories of integrative sciences: i) Level 1 integrative sciences – Systems Ecology, and Sustainability; ii) Level 2 – holistic and trans disciplinary Science of System Ecology of Sustainability.

Both categories of integrative sciences provide a theoretical foundation for two types of conceptual frameworks for system identification across the Biosphere and Ecosphere of the Earth's Life Support Environment (natural and social), namely: a) 3D-Socio-Ecologic Systems/Human and Nature Coupled Systems (3D-SES/HNCS^{s}) and b) Holistic Nature-Social Systems (h-NSEC^{s}) which have been successively adopted as "objects" in the sustainable transition (ST).

We strongly emphasize that the succession of these types of conceptual frames is correlated with the emergence and development of the Level 1 integrative sciences focused on the "natural environment" and "social environment," and Level 2 trans-disciplinary science focused on the holistic identification of the natural and social environment.

The approach and results of long-term research carried out on the nested socio-ecological systems from our test case area, extending over about hundred thousand km^2, from Ro-LDSc, including Ro-LDW, have been analyzed following the contextual, conceptual, and operational frameworks and scientific criteria described in the previous sections, as well as the requirements for 3D – Sustainable Transition and practical support to shift to the second type of holistic sustainable transition.

The expected benefits from polderisation 5747 square kilometers of wetland ecosystems and the identification of primary exergy accumulation and human appropriation didn't allow for balancing the monetary flow associated with the physical bioeconomy of Ro-LD. It was shown that the shift in the functioning regime, reflected by the deterioration of the physical bioeconomy, brought about almost 24% reduction in TEV, which should be considered only a rough, although very useful, estimation of the cost of deterioration. The proposed feed back to policy and decision-makers, and the stakeholders' network, consists of three alternative scenarios that provide the framework for an integrated assessment of short and long-term options. However, the best choice, which is assumed to be in line with the principles of sustainability, depends on the efficient use of information and communication tools regarding the complex and large-scale interactions within and among nested Danubian socio-ecological systems in the context of climate changes and regional military conflicts. That also depends on further improvement of scientific knowledge regarding complex and long-term processes, social structure, and social capacity for public participation in the policy and decision cycle.

DOI: http://dx.doi.org/10.5772/intechopen.1014638

Author details

Angheluță Vădineanu[1]*, Anca Crăciun[2] and Radu Ștefan Vădineanu[3]

1 Research Center for Systems Ecology and Sustainability/Department of Systems Ecology and Sustainable Development, Faculty of Biology, University of Bucharest, Bucharest, Romania

2 Danube Delta National Institute for Research and Development, Tulcea, Romania

3 National Centre for Sustainable Development, Bucharest, Romania

*Address all correspondence to: angheluta.vadineanu@g.unibuc.ro

References

[1] Berkes F, Folke C. Linking Social and Ecological Systems for Resilience and Sustainability. In: Berkes F, Folke C, Colding J, editors. Linking Social and Ecological Systems: Management Practices and Social Mechanisms for Building Resilience. New York: Cambridge University Press; 1998. p. 27

[2] Vădineanu A. Decision making and decision support systems for balancing Socio-Economic and Natural Capital Development. Observatorio Medioambiental. 2001;**4**:19–47

[3] Costanza R, Daly H. National Capital and sustainable development. Conservation Biology. 1992;**6**:37–46. DOI: 10.1046/j.1523-1739.1992-61.0037.x

[4] Arrow K, Bolin B, Constanza R, Dasgupta P, Folke C, Holling CS, Jansson B-O, Levin S, Maler K-G, Perrings C, Pimentel D. Economic growth, carrying capacity, and the environment. Ecological Economics. 1995;**15**(2):91–95

[5] WECD. Our Common Future: The Brundtland Report, Oxford, Oxford University Press; 1987

[6] Bettencourt LMA, Kaur J. Evolution and Structure of Sustainability Science. Proceedings of the National Academy of Sciences. 2011;**108**(49):19540–19545. DOI: 10.1073/pnas.1102712108

[7] Vădineanu A. Dezvoltarea Durabilă: Teorie și practică. Vol. **1** București. Editura Universității București Ars Docendi; 1998. p. 247

[8] Vădineanu A. Asupra unor cerințe de ordin conceptual, strategic și operațional pentru orientarea, adaptarea și eficientizarea implementării Strategiei Naționale pentru Dezvoltare Durabilă și Planului de acțiune 2030 . In: M O. editor. Dezvoltarea Durabilă a României în Context European; București; UEFISCDI Publishing House; 2023. p. 168–193

[9] Rudbeck Jespen M, Kuemmerle T, Mueller D, Erb K, Vernurg PH, Haberl H, Vesterager JP, Andric M, Antrop M, Austrheim G, Bjorn I, Bondeau A, Buergi M, Bryson J, Caspar G, Cassar LF, Conrad E, Chromy P, Daugirdas V, Van Eetvelde V, Elena-Rosello R, et al. Transitions in European Land Management Regimes between 1800 and 2010. Land Use Policy. 2015;**49**:53–64. DOI: 10.1016/j.landusepol.2025.07.003

[10] Stenzel F, Ben Uri L, Braun J, Breier J, Erb K, Gerten D, Haberl H, Matej S, Milo R, Ostberg S, Rockstrom J, Roux N, Schaphoff S, Lucht W. Breaching planetary boundaries: Over half of global land areas suffers critical losses in functional biosphere integrity. One Earth. 2025;**8**(8):101393. DOI: 10.1016/j.oneear.2025.101393

[11] Rounsevell MDA, Pedroli B, Erb K-H, Gramberger M, Busck AG, Haberl H, Kristensen S, Kuemmerle T, Lavorel S, Lindner M, Lotze-Campen H, Metzger MJ, Murray-Rust D, Popp A, Pérez-Soba M, Reenberg A, Vădineanu A, Verburg PH, Wolfslehner B. Challenges for land system science. Land Use Policy. 2012;**29**(4):899–910. DOI: 10.1016/j.landusepol.2012.01.007

[12] Haberl H, Winiwarter V, Andersson K, Ayres RU, Boone C, Castillo A Cunfer G, Fischer-Kowalski

M, Freudenburg WR, Furman E, Kaufmann R, Krausmann F, Langthaler E, Lotze-Campen H, Mirtl M, Redman CL, Reenberg A, Wardell A, Warr B, Zechmeister H. From LTER to LTSER: Conceptualizing the Socioeconomic Dimension of Long-term Socioecological Research. Ecology and Society. 2006;**11**(2):13

[13] Montgomery WC. Environmental Geology -Seventh Edition. Mc Grow New York: Hill Companies Inc; 2006

[14] Bonan GB. Ecological Climatology: Concepts and Applications. Cambridge, UK: Cambridge University Press; 2002

[15] Ramanathan VJ, Cicerone RJ, Singh HB, Kiehl JT. Trace gas trends and their potential role in climate change. Journal of Geophysical Research: Atmospheres. 1985;**90**(D3):5547–5566. DOI: 10.1029/JD090iD03p05547

[16] Roberts N. The Changing Global Environment. United Kingdom, Oxford: Basil Blackwell; 1994. p. Xii+531

[17] Chiras DD. Environmental Science: Action for a Sustainable Future, Benjamin. Redwood City, USA: Cummings Publ; 1991

[18] Lovelock JE. Bulletin of the American Meteorological Society. Vol. **67**. Boston, MA: American Meteorological Society; 1986. p. 392–397

[19] Strahler NA, Strahler HA. Introduction to Environmental Science. Santa Barbara, California: Hamilton Publishing Company; 1974

[20] Fischer-Kowalski M, Haberl H. Socioecological Transitions and Global Change: Trajectories of Social Metabolism and Land Use. Cheltenham, Gloucestershire, UK: Edward Elgar Publishing; 2007. p. 263

[21] Odum EP. Ecology: A Bridge Between Science and Society Sunderland, MA: Sinauer Associates Inc; 1997

[22] Vitousek PM, Mooney HA, Lubchenco J, J.m M. Human domination of earth's ecosystems. Science. 1997;**277**:494

[23] Ehrlich PR, Mooney HA. Extinction, substitution, ecosystem services. Bioscience. 1983;**33**:248–254. DOI: 10.2307/1309037

[24] Botnariuc N, Vădineanu A. Ecologie. Editura Didactică i Pedagogică. 1982; Bucureşti: 438

[25] Steffen W, Crutzen PJ, McNeill JR. The anthropocene: Are humans now overwhelming the great forces of nature? AMBIO: A Journal of the Human Environment. 2007;**36**(8): 614–621

[26] Di Castri F. Ecology in a context of economic globalization. BioScience. 2000;**50**(4):321–332

[27] Crutzen PL, Stoermer EF. The "Anthropocene". Global Change Newsletter. 2000;**41**:17–18

[28] Crutzen PJ. Geology of mankind. Nature. 2002;**415**:23. DOI: 10.1038/415023a

[29] Vădineanu RȘ. Teză de Doctorat "Metode și Indicatori pentru Evaluarea Capitalului Natural și a Sustenabilității Sistemelor Socio-Economice". www.rei.gov.ro/teze doctorat; București; Universitatea din București; 2008

[30] Steffen W, Sanderson RA, Tyson DT, Jäger J, Matson PA, Moore III B, Oldfield F, Richardson K, Schellnhuber HJ, Turner BL, Wasson RJ. Global Change and the Earth System: A Planet Under Pressure. In: (Global Change-The IGBP Book Series/ International Geosphere-Biosphere Programme). Heidelberg: Springer; 2004. p. 336

[31] Galloway JN, Townsend AR, Erisman JW, Bekunda M, Cai Z, Freney JR, Martinelli LA, Seitzinger SP, Sutton MA. Transformation of the nitrogen cycle: Recent trends, questions, and potential solutions. Science. 2008;**320**(5878):889–892. DOI: 10.1126/science.1136674

[32] Costanza R, Graumlich LJ, Steffen W. Sustainability or collapse? An integrated history and future of people on earth. Dahlem Workshop Reports. 2006;**96**:495

[33] Barnosky A, Hadly E, Bascompte J, et al. Approaching a state shift in Earth's biosphere. Nature. 2012;**486**:52–58

[34] Hönisch B, Ridgwell A, Schmidt DN, Thomas E, Gibbs SJ, Sluijs A, Zeebe R, Kump L, Martindale RC, Greene SE, Kiessling W, Ries J, Zachos JC, Royer DL, Barker S, T. m. Jr M, Moyer R, Pelejero C, Ziveri P, Foster GL, Williams B. The geological record of ocean acidification. Science. 2012;**335**(6072):1058–1063. DOI: 10.1126/science.1208277

[35] Pongratz J, Reick C, Raddatz T, Claussen M. A reconstruction of global agricultural areas and land cover for the last millennium. Global Biogeochemical Cycles. 2008;**22**. DOI: 10.1029/2007gb003153

[36] Crona B, Folke C, Galaz V. The anthropocene reality of financial risk. One Earth. 2021;**4**(5):618–628

[37] Turnhout E, McElwee P, Chiroleu-Assouline M, Clapp J, Isenhour C, Kelemen E, Jackson T, Miller DC, Rusch GM, Spangenberg JH, Waldron A. Enabling transformative economic change in the post-2020 biodiversity agenda. Conservation Letters. 2021;**14**: e12805. DOI: 10.1111/conl.12805

[38] Steffen W, Broadgate W, Deutsch L, Gaffney O, Ludwig C. The trajectory of the Anthropocene: The great acceleration. The Anthropocene Review. 2015;**2**:81–98. DOI: 10.1177/2053019614564785

[39] Folke C, Biggs R, V. NA, Reyers B, Rockström J. Social-ecological resilience and biosphere-based sustainability science. Ecology and Society. 2016;**21**(3). Available from: https://www.jstor.org/stable/26269981

[40] Musters CJM, de Graaf HJ, ter Keurs WJ. Defining socio-environmental systems for sustainable development. Ecological Economics. 1998;**26**(3):243–258. DOI: 10.1016/S0921-8009(97)00104-3

[41] Odum EP. Ecology and Our Endangered Life-Support Systems. Sunderland, USA: Sinauer Assoc. Inc. Publ; 1993

[42] UNCED/Rio/1992, Rio Declaration on Environment and Development

[43] Colloff MJ, Lavorel S, van Kerkhoff LE, Wyborn CA, Fazey I, Gorddard R, Mace GM, Foden WB, Dunlop M, Prentice IC, Crowley J, Leadley P, Degeorges P. Transforming conservation science and practice for

a postnormal world. Conservation Biology. 2017;**31**(5):1008–1017. DOI: 10.1111/cobi.12912

[44] Vădineanu A. Sustainable Development: Theory and Practice regarding the Transition of Socio-economic Systems Towards Sustainability, Studies on Science and Culture. Bucharest: UNESCO-CEPES; 2001

[45] Hibbard KA, Crutzen PJ, Lambin EF, N.j. DL, Mantua JR, McNeill B, Messerli, Steffen W. Decadal interactions of humans and the environment. In Integrated history and future of people on earth. In Costanza R, Graumlich.L. SW, editors. Dahlem Workshop Report. Vol. **96**. Boston, MA: MIT Press; 2006. p. 341–375

[46] Vădineanu A. Decision support systems for balancing Socio-Economic system and Natural Capital (CN↔SSE) Development. In: Proceedings of European Conference - Globalisation, Ecology, Economy Bridging Worlds, Tilburg, Netherlands. 1999. Available from: https://www.ecnc.org/publications/technicalreports

[47] Sterner T, Barbier EB, Bateman I, et al. Policy design for the Anthropocene. Nature Sustainability. 2029;**2**:14–21

[48] Sardar Z. Welcome to postnormal times. Futures. 2010;**42**(5):435–444. DOI: 10.1016/j.futures.2009.11.028

[49] Helbing D. Globally networked risks and how to respond. Nature. 2013;**497**:51–59. DOI: 10.1038/nature12047

[50] Rockström J, Steffen W, Noone K, et al. A safe operating space for humanity. Nature. 2009;**461**:472–475

[51] Rockström J, Gupta J, Qin D, Lade SJ, Abrams JF, Andersen LS, D. i. AM, Bai X, Bala G, Bunn SE, et al. Safe and just Earth system boundaries. Nature. 2023;**619**:102–111

[52] Funtowicz SO, Ravetz JR. The worth of a songbird: Ecological economics as a post-normal science. Ecological Economics. 1994;**10**(3):197–207

[53] Herbert AS. The Organization of Complex Systems. In: Pattee HH, editor. Hierarchy Theory – The Challenge of Complex Systems. New York: G. Braziller; 1973. p. 1–27

[54] Halfon E. Theory in Ecosystem Analysis. In: HALFON E, editor. Theoretical Systems Ecology. New York: Academic Press; 1979. p. 1–12

[55] Lovelock JE. Gaia: A New Look at Life on Earth. Oxford: University Press; 1979

[56] Odum EP. Fundamentals of Ecology. Philadelphia and Toronto: W. B. Saunders; 1971. 1–574 p

[57] O`Neill RV, Deangelis DL, Waide JB, Allen TFH. A Hierarchical Concept of Ecosystems. Princeton: University Press; 1986

[58] Odum HT. Systems Ecology: An introduction. New York: Wiley-Interscience; 1983

[59] Jørgensen SE, Fath BD, Bastianoni S, Marques JC, Muller F, Nielsen SN, Patten BD, Tiezzi E Ulanowicz RE. A New Ecology: Systems Perspective Amsterdam. 1 ed. Elsevier; 2007

[60] Bernard CP. Systems Analysis and Simulation in Ecology. London and New York: Academic Press; 1971. p. 601

[61] Mitsch WJ, Mitsch RH, Turner RE. Wetlands of the old and new worlds: Ecology and Management. In Mitsch WJ, editor. Global Wetlands: Old World and New. Amsterdam and Oxford: Elsevier Scientific Publishing; 1994. p. 3–56

[62] Ravetz JR. Post-Normal Science and the complexity of transitions towards sustainability. Ecological Complexity. 2006;**3**(4):275–284. DOI: 10.1016/j.ecocom.2007.02.001

[63] Holling CS. Understanding the Complexity of Economic, Ecological, and Social Systems. Ecosystems. 2001;**4**: 390–405

[64] Gunderson LS, Holling CS. Panarchy: Understanding Transformations in Human and Natural Systems. Washington D.C: Island Press; 2002

[65] Walker B, Holling CS, Carpenter SR, Kinzig A. Resilience, adaptability and transformability in social–ecological systems. Ecology and Society. 2004;**9**(2):5

[66] Walker B, Meyers JA. Thresholds in ecological and social–ecological systems: A developing database. Ecology and Society. 2004;**9**(2):3. Available from: https://www.ecologyandsociety.org/vol9/iss2/art3/

[67] Collins SL, Scott SM, Anderson CW, Gragson T, Grimm NB, Grove M, Knapp AK, Kofinas G, Magnuson J, McDowell B, Melack J, Moore J, Ogden L, Reichman OJ, Robertson GP, Smith MD, Whitmer A. Integrated Science for Society and Environment: A Strategic Research Initiative. In: Publication #23 of the US-LTER Network. LTER Network Office. Albuquerque NM: University of New Mexico; 2007. p. 35

[68] Liu J, Dietz T, Carpenter SR, Folke C, Alberti M, Redman CL, Schneider HS, Ostrom E, Pell AN, Lubchenco J, Taylor WW, Ouyang Z, Deadman P, Kratz T, Provencher W. Coupled Human and Natural Systems. AMBIO: A Journal of the Human Environment. 2007;**36**(8):639–649. DOI: 10.1579/0044-7447(2007)36[639:chans]2.0.co;2

[69] Jansson AM, Hammer M, Folke C, Costanza R, editors. Investing in Natural Capital: The Ecological economics Approach to Sustainability. Washington D.C: Island Press; 1994. p. 520

[70] Clark WC. Sustainability Science: A room of its own. Proceedings of the National Academy of Science. 2007;**104** (6):1737–1738. DOI: 10.1073/pnas.0611291104

[71] Clark WC, Dickson NM. Sustainability science: The emerging research program. Proceedings of the National Academy of Sciences. 2003;**100**(14):8059–8061. DOI: 10.1073/pnas.1231333100

[72] Vădineanu A, et al. Conceptual and operational foundation of the European Postgraduate Curriculum of Ecotechnie (3D-Sustainability)EU-Tempus: MJEP-09536/95. Background paper (669). Bucharest: European Training Foundation, Bucharest University Press; 1997. p. 73

[73] Daily G, Ruckelshous. 25 years of Valuing Ecosystems indecisions. Nature. 2022;**606**:465–466

[74] de Vos A, Quinlan A, Biggs R, Bennett EM, Martín-López B, Norström AV, Peterson GD, Schoon M, Allen CR, Andersson E, et al. Welcome home! Introducing SocSES: A society for

inclusive and impactful social-ecological research. Ecology and Society. 2025;**30** (2):32. DOI: 10.5751/ES-16164-300232'

[75] Decamps H, G PE. 1995. European River margins as indicator of Global Changes. Final Report. Environment and Climate Program of DGXII

[76] Mitch WJ, Gosselink JG. Wetlands. New York, NY: Van Nostrand Reinhold Co; 1986. p. 539

[77] Pinay G, Decamps H, Chauvet E, Fustec E. Functions of ecotones in fluvial systems. In: Naiman RJ, Decamps H, editors. The Ecology and Management of Aquatic-Terrestrial Ecotones. UNESCO, Paris; Parthenon Publishing. 1990. p. 142–169

[78] Antipa G. Regiunea Inundabilă a Dunării. Starea Ei Actuală Şi Mijoacele de a O Pune În Valoare. Bucureşti: Institutul de Arte Grafice Carol Göbl; 1910. p. 317

[79] Behr O, Gutknecht D, Modev S, Bondar C, Prohaska S. Danube river morphology: A common initiative was taken. In: Danube Watch. Vol. **1**. Vienna: ICDPR; 2000. p. 6–8

[80] Buttler A, Petts G. ERMAS II: European river margins-Role of biodiversity in the functioning of riparian systems. Bruxelles: Envronment and Climate Programme DGXII; 1995

[81] Shields MD, Deng FJ, Kato Y. The design and effects of control systems: Tests of direct- and indirect-effects models Accounting. Accounting, Organizations and Society. 2000;**25** (2):185–202

[82] Gurnell AM, Angold P, Gregory KJ. Classification of river corridors: Issues to be addressed in developing an operational methodology. Aquatic Conservation: Marine and Freshwater Ecosystems. 1994;**4**:219–231

[83] Vadineanu A, Cristofor S, Ignat G, Romanca G, Ciubuc C, Florescu C. Changes and opportunities for integrated management of the Razim-Sinoe Lagoon System. International Journal of Salt Lake Research. 1997;**6**:135–144

[84] Wachs B. Effects of water engineering measures on the structure of biocenoses and on the content of noxious pollutants in Danube. Z. Wasser – Abwasser – Forsch. 1990;**23**:173–179

[85] Arrow K, Bolin B, Costanza R, Dasgupta P, Folke C, Holling CS, B.-o. J, Levin S, K.-g. M, Perrings C, Pimentel D. Economic growth, carrying capacity, and the environment. Ecological Economics. 1995;**15**(2):91–97. DOI: 10.1016/0921-8009(95)00059-3

[86] Vadineanu A, Cristofor S, Ignat G. Phytoplankton and submerged macrophytes in the aquatic ecosystems of the Danube Delta during the last decade. Hydrobiologia. 1992;**243** (244):141–146

[87] Turner RK, Bateman IJ, Adger WN. Economics of Coastal and Water Resources: Valuing Environmental Functions. Dordrecht, The Netherlands: Kluwer Academic Publishers; 2001. p. 342

[88] Amblard. L, Colin JP. Rivers tenancy in Romania: Actors's rationales and equity outcomes. Land Use Policy. 2009;**26**(3):828–836

[89] OpenNESS - Operationalisation of natural capital and ecosystem services:

From concepts to real-world applications, 2012-2017 EU- Grant 308428

[90] VOLANTE – VISION OF LAND use Transitions in Europe EU 2010-265104

[91] RIVFunction Integrating Ecosystem Function into River Quality Assessment and Management. EU – EVKI-CT-2001-00088

[92] AQUAMONEY. Assessment of the Environmental and Resources Costs and Benefits of Water Services. EU 2006-022723

[93] Zinevici V, Parpală L, 2000. Structure and main trophic connections of the zooplankton in the large DanubeDelta lakes. IN: Proceedings of the Institute of Biology, vol. **III**, pp. 41–51. Bucureşti

[94] Zinevici V, Parpală L. The zooplankton structure and productivity in Danube Delta lacustrine ecosystems In: Tudorancea C, Tudorancea MM, editors. Danube Delta Genesis and Biodiversity, Leiden. The Netherlands: Backhuys Publisher; 2006. p. 177–210

[95] Cazacu C. Teză de Doctorat "Dinamica Structurii și Rolului Funcțional al Zooplanctonului din Ecosistemele Acvatice ale Zonei Inundabile Dunărea Inferioară". București: Universitatea din București; 2006. www.rei.gov.ro/teze-doctorat

[96] Adamescu CM. Teză de Doctorat "Dinamica structurii și rolului funcțional al fitoplanctonului din ecosistemele acvatice ale zonei inundabile/Dunărea inferioară". București: Universitatea din București; 2004; p. 221. www.rei.gov.ro/teze-doctorat

[97] Cristofor S, Vădineanu A, Sârbu A, Postolache C, Dobre R, M.c A. Long-term changes of submerged macrophytes in the Lower Danube Wetlands System. Hydrobiologia. 2003;**506**(509):625–634

[98] Vădineanu, et al. Structural and functional changes in the benthic communities from Danube Delta lakes. Verh.Internal.Verein.Limnol. 2000;**27**:2571–2576

[99] Rîşnoveanu G, Vădineanu A. Long term functional changes within the oligochaeta communities in the Danube Delta. Hydrobiologia. 2003;**506-509**:399–405

[100] Staraş M, Năvodaru I. Schimbări în structura ihtiofaunei ca efect al modificării caracteristicilor biotopului . Analele Ştiinţifice ale Institutului Delta Dunării Tulcea. 1995;**4**(1):233–239

[101] Năvodaru I, Buijse AD, Staraş M. Effects ofhydrology and water quality on the fish communityin Danube delta lakes. International Review of Hydrobiology. 2002;**87**:329–348

[102] Năvodaru I, Kiss JB, Cernișencu I. Fichery and piscivorous birds forced to sustain together in Danube Delta, Romania (Review). In: Studii și Cercetări, Biologie. Vol. **8**. Bacau, Romania. Universitatea din Bacău; 2004. p. 128–139

[103] Postolache C. The chemistry of the Danube Delta. In: Tudorancea C, Tudorancea MM, editors. Danube Delta, Genesis and Biodiversity. Leiden, The Netherlands: Backhuys Publisher BV; 2006. p. 440

[104] Iordache V. Preliminary assessment of the pollution risk associated with the restoration of the

Danube floodplain. In: Proceedings of the 5th International Symposium on Metal Elements in Environment, Medicine and Biology; 2002. p. 211–219

[105] Botnariuc N, Beldescu S. Monograph of the lake complex Crapina–Jijila. I. Fiziografie. Hidrobiologia (București). 1961;**11**: 161–242

[106] Wachs. Effects of water engineering messures on the structure of biocenoses and on the content of noxious pollutants in Danube. Z. Wasser – Abwasser – Forsch. 1990;**23**:173–179

[107] Gâştescu P. The Danube Delta: Geographical characteristics and ecological recovery. GeoJournal. 1993;**29**(1):57–67

[108] Panin N. Global Changes, Sea Level Rising and the Danube Delta: Risks and Responses. Geo-EcoMarina. 1998;**4**:19–30

[109] Bacalbașa-Dobrovici N. The Danube River and its fisheries. Special Publication, Canadian Fisheries and Aquatic Sciences. 1989;**106**:455–468

[110] Vădineanu A. Arguments in favor of integration six Romanian -LTSER sites in the global ILTER network. In: ILTER Coordinating Committee Meeting; Seattle, USA. 2003

[111] Vădineanu A. Ecosystem and Adaptive Management of Socio-Ecological Systems: Lecture Notes, Department of Systems Ecology and Sustainability. București, Romania: University of Bucharest; 2005. www.ecologie.unibuc.ro

[112] Geamana N, Vădineanu. A. Policies, Governance and Landscape change – Document analysis of EU policies implementation in Romania, working paper, VOLANTE–ENV–2010–265104. 2012

[113] Dumitru M. Romania- Country Case Study-An Integrating Land Issues into the Broader Development Agenda. Report: World Bank Bucharest. 2002

Chapter 5

Integrated Land-Use Strategies for Flood, Drought, and Water Table Management in Arid and Semi-Arid Regions

Anwarelsadat Elmahal, Wifag Hassan Mahmoud, Ahmed Abdalla, Etaib Saeed Ganawa and Abdelrahim Salih

Abstract

Dryland regions in developing countries, such as those across the Sahel and Sub-Saharan Africa, are increasingly vulnerable to complex and interlinked water-related hazards. While often characterized by water scarcity and prolonged droughts, these regions paradoxically also face recurring flash floods and riverine inundations, especially during short but intense rainy seasons. In Sudan, for instance, seasonal floods along the Nile and its tributaries have become more severe, while at the same time, localized groundwater rise, commonly referred to as water table uprising, has threatened agricultural land, infrastructure, and housing in poorly planned urban areas. This chapter examines how strategic land-use planning can mitigate these diverse and escalating risks in dryland environments. It advocates for integrated and adaptive planning approaches that recognize the dual threat of flood and drought, and the growing strain on water resources. Using tools such as Geographic Information Systems (GIS), remote sensing, hydrological modeling, and hazard mapping, planners can identify high-risk zones, improve drainage capacity, and promote resilient development. The chapter also highlights nature-based and community-led strategies, including catchment reforestation, floodplain zoning, controlled urban expansion, aquifer recharge zones, and wetland conservation. Drawing on case studies from Sudan and other dryland countries, the chapter emphasizes the importance of robust policy frameworks, institutional coordination, and inclusive governance in enabling sustainable land-use practices. Ultimately, the chapter offers a roadmap for planners, decision-makers, and researchers seeking to strengthen climate resilience and manage water extremes, both scarcity and excess, through land-use planning in some of the world's most environmentally fragile regions.

Keywords: dryland regions, water-related hazards, suistainability, geoinformatics, land use, strategic planning, policy frameworks

1. Introduction

Arid and semi-arid regions, classified as drylands based on the United Nations Convention to Combat Desertification's Aridity Index (AI ≤0.65), are among the most ecologically and socioeconomically vulnerable landscapes globally, supporting over two billion people under increasingly volatile hydrological and climatic conditions [1]. Sub-Saharan Africa (SSA) exemplifies this fragility, particularly in the Sahel region, where prolonged droughts, intense flash floods, groundwater depletion, and erratic rainfall patterns threaten water security, agriculture, and livelihoods [2–4]. These stressors are not solely climate-driven. Misaligned land-use planning, fragmented policy frameworks, and unchecked urban and agricultural expansion exacerbate ecosystem degradation and hydrological imbalances [5]. Such practices disrupt natural recharge cycles, diminish vegetation cover, and increase susceptibility to both floods and droughts. The paradox of drylands, experiencing both severe droughts and destructive floods, makes sustainable land and water management a critical imperative [6, 7].

This chapter employs a qualitative approach, structured around a comprehensive review and synthesis of scholarly literature, policy frameworks, and institutional evaluations. This methodological approach enables a critical examination of how land-use planning intersects with water and environmental management frameworks across SSA. To contextualize this analysis, Sudan is used as a central geographic case. Its diverse ecological zones reflect the broader environmental variability of the SSA region, making it a representative and analytically rich setting for exploring integrated resource management strategies. Sudan faces significant environmental and climatic pressures, especially in areas that are ecologically vulnerable. Regions like Darfur, along with the Northern, Central, and Eastern parts of the country, are particularly affected by these challenges, which compound existing fragilities and demand targeted resilience efforts. This makes it a pertinent context for examining integrated land and water management frameworks. Historic droughts (e.g., the 1980s) displaced communities into customary land systems such as the hakura in Darfur, sparking disputes over territorial rights amidst tenure fragmentation [8]. More recent events, including the 2024 floods, underscore the urgent need for coordinated adaptation strategies [9, 10]. The chapter places particular emphasis on Sudanese case studies to unpack the broader dynamics of land and water governance in dryland settings. These cases expose a complex interplay of challenges, including recurring droughts, floods, declining groundwater reserves, rising soil moisture levels, and limited adoption of geospatial technologies. Together, they point to persistent institutional fragmentation and siloed resource management. In addressing these issues, the chapter presents integrated planning approaches rooted in IWRM principles, community-based governance, and gender-responsive frameworks aimed at fostering inclusive and resilient outcomes.

On the one hand, fragmented tenure systems, weak institutions, and competing legal frameworks undermine land governance across SSA. Sudan provides a stark illustration of these issues, where drought-driven displacement disrupted existing land governance structures. At the same time, rapid urban growth and the spread of informal settlements have placed additional pressure on already overstretched administrative systems [8].

On the other hand, poor coordination among environmental, agricultural, and urban planning authorities continues to hinder sustainable land-use efforts. In Sudan and across SSA, such institutional silos exacerbate resource inefficiencies and impede adaptation. Gender-based land access disparities and limited rights for displaced

DOI: http://dx.doi.org/10.5772/intechopen.1012411

communities compound these issues, intensifying inequality and undermining long-term resilience and stability [11].

Ultimately, this chapter underscores the need for integrated planning to address the twin challenges of climate stress and fragmented governance in dryland contexts. By focusing on Sudan's environmental pressures and institutional gaps, it aims to identify practical solutions for sustainable land and water management across SSA.

2. Water-related hazards in drylands

Sudan's drylands are increasingly vulnerable to water-related hazards due to a balance between aridity and seasonal rainfall. While drought and water scarcity have historically shaped life in these regions, new patterns reveal a complex interplay of extremes, including seasonal flooding and rising groundwater levels. As shown in **Figure 1**, the global map of water stress and scarcity reveals that Sudan is under significant water stress. Climate change, rapid urbanization, and unsustainable land use exacerbate the frequency and severity of these hydrological shocks. This challenge encompasses four key dimensions: the causes of seasonal flooding, the impacts of drought, the risks of rising groundwater, and the interactions among these hazards that can amplify their effects.

2.1 Seasonal flooding: Causes and trends

Evidence indicates that natural disasters such as floods caused by extreme weather and climate have increased in frequency, magnitude, and severity around the world [12–14]. Several studies have documented the reasons behind the increase in floods in

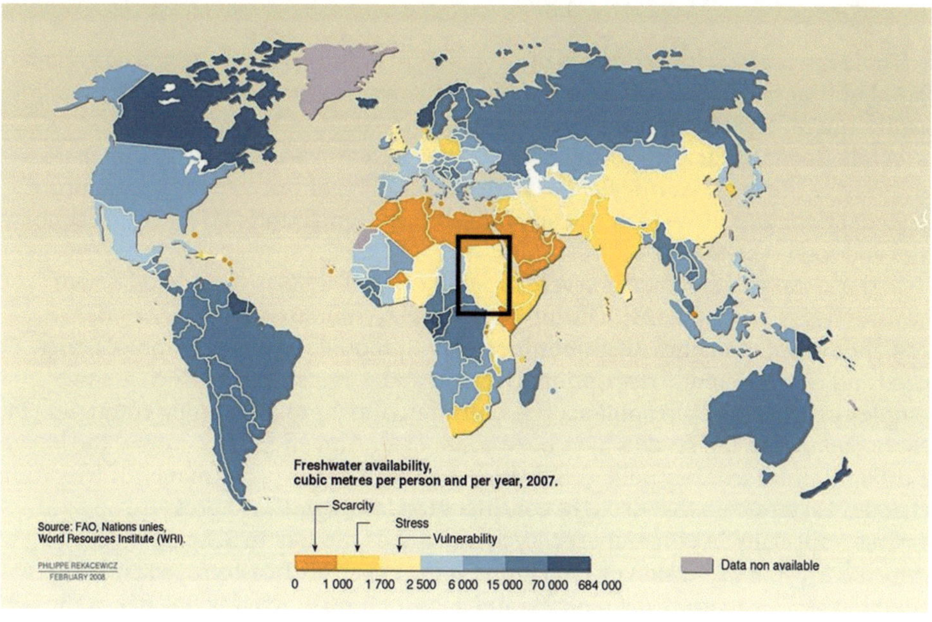

Figure 1.
Sudan's water availability status within the global water stress and scarcity. The black frame indicates the location of old Sudan before the succession of South Sudan. Source: https://www.grida.no/resources/5651.

arid and semi-arid regions such as Sudan, where seasonal flooding remains one of the most pressing water-related hazards, particularly in rapidly expanding urban areas such as Khartoum [15]. Global warming, represented by intense rainfall, sea level rise, and climate change, is among the main causes of flooding, especially flash flooding in Sudan [15]. The rainy season, compressed between July and September, produces intense upstream rainfall in the Ethiopian Highlands, contributing roughly 85% of the Nile's total flow. When the Blue Nile surges past 16.5 meters at the Khartoum gauge station, low-lying areas such as Tuti Island and parts of Omdurman are routinely at risk [13, 16].

Notably, Sudan's wadis, typically dry streambeds, can become torrents during heavy rain. In 2013, flash floods triggered by intense rainfall (approximately 40–50 mm) caused severe damage in Khartoum, particularly in Marabie El Shareef and Sharg El-Nil, where Wadi Soba overwhelmed infrastructure due to undersized culverts [15, 17]. These effects are exacerbated by unregulated urban sprawl into natural waterways. While floods have dire consequences, certain parts of Sudan, such as the Blue Nile Basin, also face prolonged droughts (1984, 2002, 2009, 2015), influenced by El Niño and La Niña events that disrupt rainfall patterns [18]. The 1984 drought remains one of the most devastating in Sudan's history.

In response to mounting risks, preparedness initiatives are gaining traction. The Eastern Nile Technical Regional Office (ENTRO), operating under the Nile Basin Initiative, introduced the Flood Preparedness and Early Warning (FPEW) system, which provides daily forecasts, risk maps, and alerts to more than 200,000 individuals [19]. Community-based measures, such as color-coded flood markers and mobile alerts, offer localized support mechanisms. These developments highlight a growing imbalance between natural forces, human expansion, and existing infrastructure. With floods and droughts intensifying under climate change, further investigation and adaptive strategies are essential to mitigate future impacts on Sudan's drylands.

2.2 Drought and water scarcity: Long-term impacts

There is not a precise definition of drought [20]. It can be defined as a sustained period of time without significant rainfall. However, for more definitions of droughts, readers can refer to **Table 1** in [20].

While floods often attract attention, it is drought and water scarcity that persistently shape life in Sudan's drylands. With the majority of the country's agriculture dependent on rainfall, even minor shifts in climate significantly affect rural economies and food systems [21].

In the history of Sudan, the severe drought of 1984–1985 is considered one of the most devastating [22, 23]. During this drought, millions of livestock perished, croplands failed, and entire communities were uprooted, resulting in long-lasting social and environmental disruption. Drought and water scarcity, which are two complex hydrological phenomena, pose substantial risks to developing countries [24, 25]. In Sudan, limited studies have been conducted. Ahmed [22] examined the impact of drought, food security policy, and climate change on the performance of irrigation schemes. He reported that drought continues to place significant pressure on food systems, especially in vulnerable ecosystems such as the Gezira Scheme. According to Ahmed [22], "the expansion of food crops at the expense of cash crops resulted in a reduction of 83% in gross net benefits and a loss of 63% in irrigation water in the GS since the severe drought of 1984."

According to the European Commission—Joint Research Center [26], during the years 2009, 2012 and 2015, Sudan has experienced only mild or short droughts

Drought Parameters	Gadarif Station *(Ave. Ann. Rainfall = 607.7 mm Stand. Dev. = 131.1 mm)*	Halfa Algadeeda Station *(Ave. Ann. Rainfall = 261.7 mm Stand. Dev. = 125 mm)*
Longest Dry Spell Duration (LDSD)	5 years (2009–2013)	11 years (1982–1992)
Drought Tendency (DT) = D*/W**	1.28	1.11
Number of Dry Spell Events (N)	6	11
Average Dry Spell Duration (ADSD) = D/N	3.83	1.91
Standard Total Cumulative Drought (STCD) = Σ D	−16.19	−14.89
Average Dry Spell Index (ADSI) = Σ D/N	−2.70 (Extreme Drought)	−1.35 (Moderate Drought)
Largest Multi-Year Drought (LMYD)	−3.50 (The successive years 1990–1992)	−10.98 (The successive years 1982–1992)
Largest Single-Year Drought (LMSD)	−2.18 (For the year 1984)	−1.80 (For the years 1984 and 1990)

**D: Dry years;*
***W: Wet years.*

Table 1.
Dry spell analysis of Gadarif and Halfa Algaddeda stations (1980–2020) in the rainfed agricultural schemes in Sahelian Sudan (Author: Wifag H. Mahmoud).

compared to the eighteens' and nineteens' strong drought events. This is demonstrated by **Figure 2**, which shows the classification of the dry and wet years according to the Standardized Precipitation Index (SPI) for two meteorological stations in Eastern Sudan, i.e., Gadarif and Halfa Algadeeda. The SPI is defined as the deviation of the annual rainfall from the annual mean. Both stations are located in the largest commercial rainfed agricultural schemes that are subject to the drought as well as the threat of increased dry spell episodes within the rainy season. For instance, the dry spell led to the failure of the 2012 agricultural season, by a 64% decrease in yield compared to that of the year 2007. **Table 1** presents the drought and dry spell analysis for the same stations for the period 1980–2020. It is clearly shown that the eightieth and ninetieth year was characterized by the largest multi and single year drought for both stations. In addition, the average dry spell index (ADSI) shows extreme drought for Gadarif (−2.7), while the longest dry spell duration (LDSD) was for Halfa Algadeeda station for 11 years (1982–1992). This analysis reflects the vulnerability of the agricultural lands in SSA to drought, which is exacerbated by climate change and variability. Such threats lead to inevitable environmental and socioeconomic implications.

In another regional study, drought conditions in North Darfur were evaluated using indices such as the Vegetation Condition Index (VCI) and Soil Moisture Content Index (SMCI), with findings indicating prolonged drought events between 2007 and 2011 [27]. Biophysical and socioeconomic indicators were used to assess vulnerability to drought risk, and the Combined Drought Index (CDI) was validated as an effective tool. These droughts resulted in reductions in crop yields, livestock fatalities, soil degradation, and vegetation loss [27, 28]. Long-term data indicate

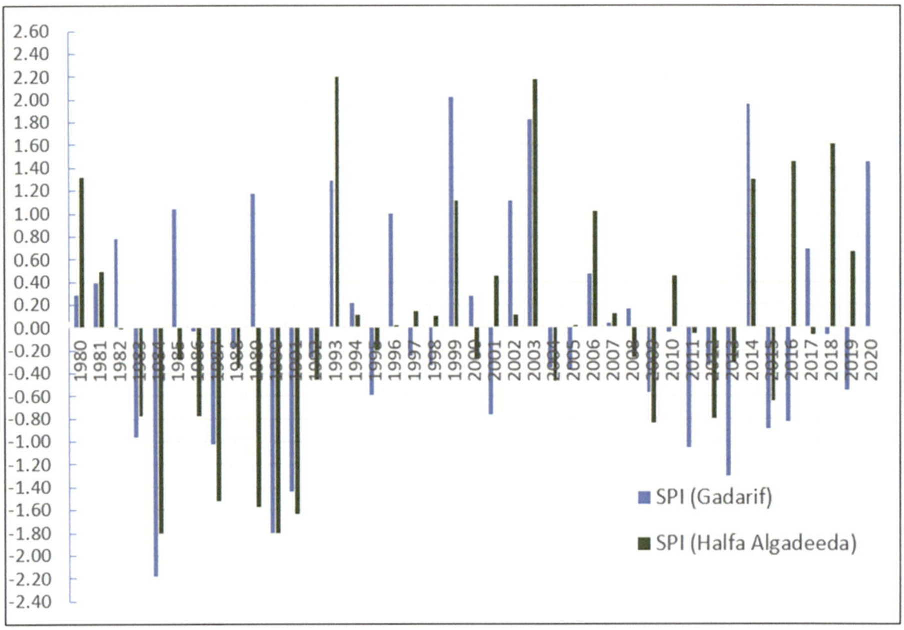

Figure 2.
Standardized precipitation index (SPI) analysis of the dry and wet years of Gadarif and Halfa Algaddeda stations (1980–2020) in the rainfed agricultural schemes in Sahelian Sudan (Author: Wifag H. Mahmoud).

significant rainfall decline since the 1970s in regions such as Darfur and Northern Kordofan [28]. Desertification continues to expand, converting once-fertile land into arid zones.

Environmental stress is further exacerbated by human-induced pressures. Deforestation, overgrazing, and unsustainable practices have stripped vegetation cover, degraded soils, and reduced the land's ability to recover from drought events. In Darfur, these stressors have contributed to resource-based conflict, illustrating how ecological degradation can threaten peace and social cohesion [29].

Climate projections indicate that northern Sudan receives less than 50 mm of annual rainfall, while southern regions may exceed 1600 mm. Yet precipitation is increasingly erratic. Rising temperatures are accelerating evapotranspiration, reducing soil moisture even during wet periods. Some projections warn of a 20% reduction in precipitation and up to a 4°C temperature rise by the end of the century [30, 31].

At the same time, Sudan's reliance on groundwater, particularly the Nubian Sandstone and Umm Ruwaba formations, is growing. These reserves are vital during drought years but are being depleted faster than they can naturally recharge [32].

To monitor the impacts of drought and water scarcity, techniques such as the Standardized Precipitation Index (SPI), Standardized Precipitation Evapotranspiration Index (SPEI), and Soil Water Deficit Index (SWDI) have been employed using station-based rainfall records.

Despite the magnitude of these challenges, viable solutions exist. Drought-tolerant crop varieties, water harvesting systems, reforestation initiatives, and improved land governance offer promising responses, provided they are supported by sufficient investment, institutional coordination, and long-term commitment [21]. Monitoring and understanding drought dynamics remains essential for mitigating risk and sustaining land and water systems.

2.3 Groundwater rise and water table uprising: A hidden threat

In many parts of the world, the primary concern regarding groundwater is depletion, where aquifers are tapped faster than they can be replenished, resulting in declining water tables and long-term scarcity. Yet, Sudan presents a paradoxical challenge: in several agricultural and urban areas, groundwater levels are not falling but rising, sometimes alarmingly so. Though seemingly counterintuitive, this trend carries significant implications for infrastructure stability, agricultural productivity, and public health, particularly in a country that also faces chronic drought in other regions.

One of the most striking examples of this issue can be found in the Gezira Scheme, Sudan's oldest and most extensive irrigation project. Once a model of agricultural development, the scheme now suffers from decades of water mismanagement. Its unlined canals leak profusely, and in the absence of adequate drainage systems, much of this excess water seeps into the soil, pushing the groundwater table closer to the surface, often within one to two meters. As a result, vast tracts of farmland suffer from waterlogging and salinization, which reduces crop yields and gradually degrades soil health [33]. These problems are exacerbated by the region's flat terrain and limited evapotranspiration during the cooler months, which reduces natural discharge pathways and allows water to accumulate in the subsurface [34].

Urban centers, such as Khartoum, are facing similar, though more complex, dynamics. Unregulated urban expansion, inefficient wastewater disposal, and inadequate stormwater drainage systems have all contributed to a steady rise in groundwater levels. This upward movement poses a twofold threat. Structurally, it comprises building foundations, particularly in neighborhoods built on clay-rich soils that swell when saturated. From a public health perspective, the shallow groundwater becomes a conduit for contamination, especially in densely populated informal settlements where pit latrines and septic tanks are often poorly maintained [34, 35].

Underlying geology further shapes this phenomenon. The Umm Ruwaba Formation, which dominates much of central Sudan, consists of fine-grained sediments with low permeability. These soils trap water near the surface, leading to perched water tables and limited vertical drainage. Meanwhile, the Nubian Sandstone Aquifer, more expansive and deeper, typically remains disconnected from modern recharge, except in localized zones such as wadis and alluvial fans where episodic floodwaters may infiltrate [35]. These geological differences mean that groundwater behavior varies widely across the country, complicating any one-size-fits-all management approach.

Rising groundwater in a drought-prone country might seem like a fortunate development. But in practice, its effects are often adverse. Water tables that approach the surface bring salt upward through capillary action, degrading soils and rendering fields unproductive. Stagnant, shallow groundwater also fosters mosquito breeding, increasing the risk of malaria and other vector-borne diseases. Without adequate drainage, infrastructure like roads and buildings becomes susceptible to subsidence and deterioration. What should serve as a lifeline in dry periods turns into a concealed danger, undermining the very development it was initially intended to support.

Sudan faces a critical issue due to the absence of a centralized groundwater monitoring system. Sudan lacks a centralized groundwater monitoring system. Most available data come from donor-funded projects or isolated studies near cities, leaving large rural areas effectively unmapped. This absence of coordinated oversight means that rising water tables often go unnoticed until they cause visible damage.

Addressing the problem will require a multi-pronged approach, modernizing irrigation systems to reduce losses, installing controlled drainage in waterlogged zones, and empowering communities with the tools and training to monitor local water conditions. Without such measures, groundwater rise will continue to be an insidious threat, quietly destabilizing lives, livelihoods, and landscapes from below.

2.4 Interactions between these hazards

The real challenge for Sudan may not lie in individual water-related hazards, but in how these hazards interact, amplify, and cascade into one another. Drought, flooding, and groundwater shifts are no longer isolated events; they are increasingly intertwined, feeding off each other in ways that make conventional, single-hazard responses inadequate. A telling example of this complexity is the phenomenon of post-drought flooding. After extended dry periods, soil becomes hardened and cracked, stripped of vegetative cover and organic matter. These conditions drastically reduce infiltration capacity. When heavy rainfall eventually returns, the parched ground is unable to absorb the water, resulting in rapid surface runoff and flash floods. Such events have become more frequent and intense across the Blue Nile Basin, where communities now face increasingly unpredictable cycles of drought followed by deluge.

The impacts of these floods extend beyond the surface. In areas underlain by fractured bedrock or permeable sediments, floodwaters can infiltrate quickly and recharge shallow aquifers. While this may temporarily ease water scarcity, it also introduces new risks. A recent study in East Africa showed that such recharge episodes can raise groundwater tables sharply within weeks, potentially causing waterlogging, rising salinity, and infrastructure challenges in irrigated or low-lying regions [16]. At the same time, not all aquifers respond so readily. GRACE satellite data have revealed that many groundwater systems in drylands, including those in Sudan, experience long delays in recovery following prolonged drought. Even after rainfall returns, it can take years for these aquifers to replenish, leaving dependent communities exposed to water stress long after the drought has officially ended [36]. In support of this, GRACE-derived equivalent water height (EWH) measurements over areas such as the Dinder and Rahad basins in Eastern Sudan display strong seasonal fluctuations and long-term trends that reflect both episodic recharge and delayed groundwater responses. These variations, observed across different GRACE processing centers (CSR, JPL, and GFZ), confirm that terrestrial water storage in Sudan is shaped by a complex balance between surface inputs and subsurface inertia, underscoring the need to account for both in hazard planning [37]. This dynamic interaction of hazards has severe human consequences. Livestock losses during drought are often followed by crop destruction during floods. Homes that survive one extreme may not survive the next. In rural areas near seasonal rivers and depressions, these repeated shocks create a cycle of loss and vulnerability that is difficult to break. Yet, despite these realities, hazard management in Sudan remains fragmented. Flood response falls under one agency, drought under another, and groundwater governance is often neglected entirely. Forecasting tools and planning frameworks rarely integrate across these domains, limiting the country's ability to anticipate and mitigate cascading risks.

To move forward, Sudan urgently needs to adopt a multi-hazard approach. Early warning systems must reflect the connections between surface and subsurface water processes. Local strategies from improved drainage to adaptive cropping calendars must be designed with the recognition that these hazards do not operate in isolation. Ultimately, understanding and managing the interactions between water-related hazards may prove to be the most decisive factor in building long-term resilience for Sudan's dryland communities.

3. The role of land-use planning in risk mitigation

Land-use planning plays a critical role in risk mitigation by guiding where and how land is developed, helping to reduce vulnerability to natural hazards. In drylands, effective planning can limit settlement and infrastructure in flood-prone areas, promote sustainable water management, and support land practices that reduce the impact of droughts, such as preserving vegetation cover and enhancing soil retention. This is especially vital in SSA's drylands, where climate variability heightens exposure to both floods and droughts.

3.1 Land use in the drylands of Sub-Saharan Africa

The drylands of Sub-Saharan Africa (SSA) encompass arid, semi-arid, and dry sub-humid zones. Land-use patterns in these regions are fundamentally shaped by the scarcity and variability of water, fragile ecosystems, and prevalent socioeconomic pressures [38]. These factors collectively render drylands hotspots for natural disasters, poverty, and conflict. The primary land-use categories observed in SSA drylands include deserts and bare lands, rangelands, croplands, forests and woodlands, and built/urban areas.

Deserts and Bare Lands The desert and bare land category encompasses areas with very little to no vegetation, including natural desert landscapes, rocky outcrops, sand dunes, and severely degraded lands. These areas are often characterized by severe soil erosion, low soil fertility, and limited biological activity, representing the ultimate consequence of desertification [30]. Both human activities and severe environmental conditions render these areas unsuitable for productive use and contribute to dust storms and overall environmental deterioration.

Rangelands constitute the most extensive land-use type in SSA drylands. These vast areas of natural grasslands, savannas, and shrublands are primarily utilized for the extensive grazing of livestock. They are central to pastoralism and agro-pastoralism activities, supporting the livelihoods of millions of people and significantly contributing to national economies [39]. Pastoralism traditionally involves various forms of mobility, allowing herders to follow the seasonal availability of pasture and water [40]. Agro-pastoralism, on the other hand, combines livestock keeping with opportunistic rainfed cultivation. However, these vital systems are increasingly vulnerable. Reduced mobility, often exacerbated by climate change impacts, leads to overgrazing, soil compaction, erosion, and the loss of valuable fodder species [41]. Furthermore, escalating population growth and food demand intensify competition with expanding croplands for diminishing grazing areas and increasingly scarce land and water resources, often escalating into conflict [41]. This dynamic interplay and overlap between land-use categories underscore the complex challenges faced by these regions.

Croplands: The cropland category in SSA drylands comprises both rainfed and irrigated agriculture. Rainfed agriculture areas are dedicated to growing crops, ranging from smallholder subsistence plots to larger, sometimes mechanized, commercial farms. This form of agriculture is crucial for food security, particularly for the large rural populations [42]. Common crops include drought-tolerant grains such as millet, sorghum, and maize, as well as various legumes. As these areas rely solely on natural rainfall, they are highly susceptible to rainfall variability, frequently leading to crop failures during droughts. Moreover, increasing food demand often drives the expansion of cultivation onto marginal, easily erodible, and low-nutrient lands, frequently competing directly with rangelands. In contrast, irrigated agricultural areas receive water through artificial irrigation systems from surface or groundwater sources. While covering a smaller total area compared to rainfed lands, these areas are often highly productive and essential for more reliable food production, capable of supporting higher population densities. Irrigation systems can range from traditional flood recession practices along major rivers, such as the Nile, to large-scale modern irrigation schemes, e.g., the Mwea Irrigation Scheme in Kenya. Challenges facing this category include significant pressure on water resources due to high demand, the risk of salinization from improper irrigation, and often high energy costs for pumping.

Forests and Woodlands: Sparse to dense tree and shrub cover, adapted to arid conditions, characterize the forest and woodland land-use type. This includes ecosystems such as Acacia-Commiphora woodlands, Miombo woodlands, and riverine forests along wadis (ephemeral water courses). Beyond being a crucial source of fuelwood, charcoal, timber, and non-timber forest products (e.g., gums) for local communities, these areas are vital for rural livelihoods. They also contribute significantly to environmental stability by providing essential ecosystem services such as soil stabilization, erosion control, biodiversity habitats, and the regulation of local microclimates. However, high rates of deforestation and degradation, driven by charcoal production, fuelwood collection, agricultural expansion, and unsustainable logging, further contribute to widespread land degradation.

Built/Urban Areas: While occupying a relatively small percentage of the total dryland area, urban and built-up areas are centers of economic activity, services, and population concentration. This land use is rapidly expanding, frequently encroaching upon surrounding agricultural or natural land. Such unplanned growth can lead to the establishment of informal settlements in hazard-prone areas. Moreover, it exacerbates water scarcity, increases demand for resources, generates waste, and contributes to urban heat island effects.

3.2 Land-use change in SSA drylands: Key drivers

The dynamics among land-use categories are continuously evolving, shaped by a series of key factors. (i) Population growth and demographic shifts, including rural-urban migration, place increasing demands on land resources. (ii) Climate change introduces compounding effects through droughts, floods, erratic rainfall patterns, and rising temperatures [43–45]. (iii) Agricultural expansion, both rainfed and irrigated, is driven by growing food demand. (iv) Unsustainable resource management practices, such as overgrazing, deforestation, and unsuitable farming techniques, contribute to environmental degradation [43]. (v) Poverty and livelihood insecurity often push communities toward exploitative land use. (vi) Weak land governance and insecure tenure systems hinder effective land management. (vii) Conflict and displacement lead populations into marginal areas, intensifying ecological strain [46].

3.3 Overarching challenges shaping land use

Land use in the SSA is a dynamic and multifaceted issue, shaped by several critical and interconnected forces. (i) Climate change poses a severe challenge, as SSA drylands are warming at twice the global rate, increasing evapotranspiration and water scarcity, and causing erratic rainfall that leads to droughts, flash floods, and declining agricultural viability [43, 44, 47]. (ii) Land degradation and desertification, driven by unsustainable practices like deforestation, overgrazing, and poor farming methods, reduce productivity and exacerbate vulnerability to climatic shocks [30, 48]. (iii) Population pressure continues to mount, prompting the expansion of agricultural activity into ecologically fragile areas and the overuse of limited resources [49]. (iv) Governance and policy gaps, particularly weak tenure systems and limited implementation capacity, obstruct effective sustainable land management [50]. (v) Poverty and food insecurity function both as causes and effects of unsustainable land use, forcing communities into short-term survival strategies that deepen environmental degradation [48]. Recent literature emphasizes not only these escalating challenges but also the need for innovative and inclusive strategies to foster sustainable land management in the region [51].

3.4 Challenges to Sustainable Land Management (SLM) adoption

The implementation of Sustainable Land Management (SLM) practices in the drylands of the SSA is constrained by several major challenges. (i) Farmers often lack sufficient knowledge and training regarding sustainable agricultural techniques, which limits effective adoption [48] (ii) Many SLM methods require considerable initial investments in labor or capital, presenting barriers for resource-constrained households. (iii) Insecure land tenure prevents long-term investment, as farmers face uncertainty about future ownership and return on effort [40]. (iv) Access to finance remains limited, with few opportunities for credit, subsidies, or other financial support to facilitate adoption. (v) The immediate need for food security compels farmers to prioritize short-term production goals over sustainable, long-term land stewardship. These factors collectively impede progress toward widespread SLM adoption and highlight the importance of targeted support, capacity building, and inclusive policy reform.

3.5 Emerging research-based solutions

Remote Sensing (RS) and Geographical Information Systems (GIS) are playing an increasingly pivotal role in monitoring land-use change, diagnosing degradation, and informing spatial planning across dryland areas in SSA. Several emerging research-driven strategies complement these technological tools. (i) The Water-Energy-Food-Environment (WEFE) Nexus promotes integrated approaches that account for trade-offs and synergies among interconnected sectors [52]. (ii) Nature-based Solutions (NbS) harness ecological processes as sustainable, cost-effective alternatives to conventional infrastructure, particularly in informal settlement upgrades [53]. (iii) The upscaling of Sustainable Land Management (SLM) emphasizes the transition from isolated pilot projects to broad-based systemic change, facilitated by financial incentives, policy support, and capacity building initiatives. (iv) Indigenous and traditional ecological knowledge is gaining recognition for its adaptability and relevance in dryland contexts. When effectively blended with scientific insight and strong governance, it contributes to integrated and risk-informed land use planning [54].

3.6 The concept of risk-informed land use planning

Risk-informed land use planning offers a proactive and integrated approach that embeds risk considerations throughout land use decision-making, from policy formulation to enforcement [55]. Its core objectives include preventing the emergence of new risks, mitigating existing ones, and fostering resilience across both communities and ecosystems. (i) The first pillar focuses on understanding and assessing risks through comprehensive identification and analysis of vulnerability, exposure, and risk dynamics using both qualitative and quantitative methods [56]. (ii) The second pillar emphasizes integrating risk into planning by aligning sectoral policies with risk insights and steering investments away from high-risk zones. (iii) The third pillar aims to strengthen resilience and adaptive capacity by reducing exposure, fortifying local systems, leveraging natural solutions, and incorporating early warning mechanisms. (iv) The fourth pillar calls for participatory and inclusive processes, involving communities, diverse stakeholders, and transparent communication frameworks. (v) The fifth pillar promotes continuous learning and adaptation through regular monitoring and evaluation, treating land-use planning as an ongoing, cyclical process responsive to changing risks.

3.7 Spatial planning and hazard zoning

Spatial planning and hazard zoning are essential and interconnected components of building resilient communities in hazard-prone drylands. Spatial planning is a strategic, multidisciplinary approach to land use and environmental development across multiple scales. Its central purpose is to determine where activities are located and how land uses interact within a defined geographic area. Key attributes of spatial planning include being holistic and integrated, forward-looking in anticipation of climate change and resource scarcity, regulatory in guiding development, participatory through stakeholder involvement, and responsive to complex issues such as climate adaptation. In the context of risk mitigation, spatial planning provides the overarching framework for incorporating hazard considerations into development, with the goal of preventing new risks and guiding risk reduction efforts [57]. In contrast, hazard zoning—also referred to as hazard mapping or risk zoning—is a focused, operational tool nested within the spatial planning process. It addresses the spatial distribution of hazards and associated risks by identifying and mapping hazard-prone areas, classifying them based on likelihood and impact, and applying targeted land-use regulations tailored to each risk category [58].

3.8 Drought-flooding paradox in drylands of SSA

The phenomenon of the drought-flood paradox is increasingly observed in SSA drylands. This involves prolonged dry spells suddenly interrupted by intense, short-duration rainfall events that lead to destructive flash floods [44]. Global warming contributes to more extreme weather patterns, intensifying this paradox. Prolonged droughts dry out soils, reduce vegetation cover, and compact the land, leading to rapid surface runoff and flash floods, even in normally arid areas, when sudden heavy rains occur. Furthermore, land degradation significantly reduces the land's infiltration capacity during rainfall, resulting in less groundwater recharge and increased surface runoff as destructive floods [43]. Flood risk is further heightened by unplanned urban expansion into wadis, floodplains, or low-lying areas, and the proliferation of impervious surfaces. Conversely, reliance on limited and vulnerable water sources makes these same settlements highly susceptible to drought [47].

3.9 Land-use planning strategies for concurrent drought and flood challenges

Land-use planning in drylands plays a foundational role in developing a cohesive framework that directs human activities toward reducing hazard exposure, enhancing resilience, and promoting sustainable resource use. Addressing the concurrent challenges of drought and flood requires a multifaceted approach rooted in integration, adaptability, and ecosystem sensitivity. (i) Water resources management entails strategies such as rain and floodwater harvesting for storage [59], facilitating groundwater recharge and sustainable extraction, managing floodplains and wadis for controlled inundation, and adopting water-sensitive urban designs to reduce runoff and improve infiltration. (ii) Land-use zoning and development control involve implementing hazard-sensitive zoning to limit development in vulnerable areas, conserving critical ecosystems like dryland forests and wetlands that serve as natural buffers, and promoting sustainable agriculture and agro-pastoral systems through techniques like Integrated Soil Fertility Management (ISFM) and agroforestry to improve soil health and water retention [60]. (iii) Integrated governance and community participation focus on fostering coordination among government sectors and stakeholders [51], incorporating indigenous and traditional knowledge into planning processes [54], designing flexible and responsive planning mechanisms for climate uncertainty, and encouraging regional collaboration in managing shared resources.

4. Applications, case studies, and practical insights

Across Sudan and much of SSA, escalating water crises stem not only from climate change, but also from poorly regulated land-use practices and rapid land cover transformations [61]. Frequent and prolonged droughts continue to devastate agriculture and rural livelihoods [43, 44, 47], while intense and erratic rainfall events trigger catastrophic floods that displace hundreds of thousands annually and inflict multi-billion dollar losses on infrastructure and ecosystems [8]. Escalating the impact of these hydrological challenges are increased soil moisture, a steady drop in groundwater availability, and widespread pollution of water resources. Together, these factors significantly limit access to clean and dependable water.

Beyond natural factors, human-driven land degradation plays a pivotal role. The expansion of urban centers, deforestation, conversion of rangelands, and unsustainable farming techniques alter infiltration rates, reduce natural water storage, and increase surface runoff. These land cover shifts amplify flood risk, accelerate erosion, and diminish the resilience of dryland ecosystems. A truly effective management strategy must therefore integrate both climatic realities and anthropogenic land change dynamics into adaptive, region-specific solutions.

Although satellite missions such as Soil Moisture Active Passive (SMAP) and Gravity Recovery and Climate Experiment (GRACE/GRACE-FO) offer transformative capabilities for monitoring terrestrial water dynamics, their practical application in Sudan and much of Sub-Saharan Africa remains limited. These technologies provide near-real-time data on soil moisture, groundwater fluctuations, and total water storage, which are essential for drought preparedness, aquifer sustainability, and flood forecasting. Yet, in Sudan, their use is largely confined to academic and pilot research, with minimal uptake in governmental water management systems, disaster response frameworks, or policy planning.

This underutilization is particularly concerning given the region's vulnerability to hydrological extremes, including prolonged droughts, flash floods, and groundwater depletion. In Sudan, where ground-based monitoring networks are sparse and hydrological data is fragmented, satellite-derived insights could fill critical gaps in understanding water availability and stress. However, barriers such as limited technical capacity, lack of institutional coordination, and insufficient investment in geospatial infrastructure have hindered the integration of these tools into national early-warning systems and land-use planning.

Across SSA, similar challenges persist. Despite the proven value of remote sensing for climate resilience, many countries have yet to embed these datasets into operational decision-making [62]. Bridging this gap requires not only technical training and infrastructure upgrades but also policy reforms that recognize the strategic importance of Earth observation data in managing water resources under increasing climate and land-use pressures.

Water crises in Sudan and Sub-Saharan Africa stem from a combination of climate variability and unsustainable land-use practices, including deforestation, urban expansion, and agricultural mismanagement. These drivers intensify droughts and floods, degrade ecosystems, and compromise access to clean water through groundwater depletion and pollution. Addressing these challenges requires integrated land-use planning that accounts for both environmental and human-induced pressures.

Despite the potential of satellite technologies like SMAP and GRACE/GRACE-FO to monitor soil moisture and groundwater storage, their operational use remains limited. In Sudan, such data is often restricted to academic research, leaving a gap in national early-warning systems and water management. Bridging this divide calls for improved technical capacity, cross-sector coordination, and policy reforms that embed Earth observation tools into decision-making processes.

5. Integrated planning and governance strategies for water resilience

Water is fundamental to life, but its availability and quality are increasingly threatened by climate change, population growth, and unsustainable management practices. In many regions, particularly drylands, these pressures manifest as escalating droughts, unpredictable floods, and diminishing freshwater supplies. Building water resilience, the capacity of a system to cope with, adapt to, and recover from water-related shocks and stresses, is paramount for sustainable development and human well-being [63]. Achieving this requires a shift from fragmented, reactive approaches to holistic, integrated planning and governance strategies that address the multifaceted nature of water challenges. This involves coordinated efforts across sectors, scales, and stakeholders, encompassing both structural and non-structural interventions.

5.1 Floodplain zoning and buffer creation

Floodplains are naturally dynamic areas, often fertile and attractive for human settlement. However, uncontrolled development in these zones can lead to catastrophic consequences during flood events.

Floodplain zoning employs a risk-based methodology for land-use planning, categorizing flood-prone regions into distinct zones according to the frequency and intensity of flooding events. A key element is the utilization of flood hazard maps, which assist in identifying regions at high, medium, or low risk, typically characterized by flood return periods such as 1-in-10, 1-in-50, or 1-in-100-year floods. In certain areas, land-use restrictions are implemented to manage growth, designating high-risk zones for non-vulnerable purposes such as agriculture, open space, or recreation, while permitting more resilient infrastructure in lower-risk zones. Flood-resilient construction approaches include raising structures above the basic flood elevation or implementing flood-proofing systems. Moreover, floodplain zoning frequently corresponds with ecosystem-based strategies, safeguarding natural floodplain services such as water retention, groundwater recharge, and habitat conservation [64].

Complementing zoning, the creation and preservation of natural buffers are critical for enhancing water resilience. Ecosystems like wetlands, forests, and natural grasslands act as sponges, absorbing excess water, slowing down runoff, and reducing flood intensity downstream. These buffers also filter pollutants, recharge groundwater, and provide vital habitats [65].

5.2 Drought-resilient agricultural and land-use practices

Drought-resilient agricultural and land-use practices encompass techniques designed to sustain agricultural output and ecological stability during water constraints. These include soil moisture conservation techniques (e.g., mulching, conservation tillage), drought-resistant crop varieties, efficient water management (e.g., drip irrigation, rainwater harvesting), agroforestry, and crop diversity [66]. Land-use planning enhances drought resistance on a broader scale by promoting sustainable rangeland management, safeguarding watersheds, and preserving vegetation that aids in water retention and microclimate regulation.

Key agricultural strategies include conservation agriculture principles such as minimum tillage, permanent soil cover, and crop rotation [66]. Water harvesting techniques like contour farming, terracing, and small-scale dams are also vital. Additionally, promoting agroforestry systems that integrate trees into farming landscapes adds ecological and economic resilience [67]. In pastoral areas like Ethiopia, community-based rangeland management has shown how adaptive grazing strategies and traditional knowledge can sustain livelihoods [68].

5.3 Groundwater recharge and regulation zones

Groundwater recharge and regulation zones are essential for the replenishment of underground aquifers. These zones are generally situated where surface water infiltrates the soil to recharge aquifers. Effective management includes limiting impervious surfaces, preventing pollution, promoting afforestation, and preserving wetlands [69].

Managed Aquifer Recharge (MAR) systems deliberately introduce water into aquifers using techniques such as infiltration basins, injection wells, and spreading grounds. MAR not only augments supply, but can improve water quality through natural filtration [70]. Regulation zones must include legal frameworks and governance structures to control abstraction, issue permits, and enforce limits, ensuring sustainable use [71].

5.4 Legal and institutional frameworks

Legal and institutional frameworks for integrated water governance direct the planning, administration, and equitable distribution of water resources. These include water legislation, regulatory systems, institutional structures, and policy tools such as pricing and permits. Effective governance ensures coordination across institutions and sectors [72].

IWRM promotes coordinated development and management of water, land, and related resources [73, 74]. Legal frameworks define rights and enforcement mechanisms, while institutions implement laws and policies. Strengthening institutional capacity and inter-agency collaboration is essential for water resilience.

5.5 Mainstreaming climate and water risks into policy

Incorporating climate and water risks into policy involves integrating climate variability, change effects, and water hazards into planning and regulation. This ensures alignment across sectors like agriculture, urban planning, and health [75].

Hydrological science, climatology, and remote sensing support water risk assessments. Science-based assessments inform legal frameworks and decision-making. Policies must include climate-resilient infrastructure, nature-based solutions, and early warning systems. Organizations like the Global Water Partnership advocate for mainstreaming adaptation into water policies [76, 77].

5.6 Community-based adaptation and traditional knowledge

Community-Based Adaptation (CBA) empowers local communities to identify, plan, and implement adaptive strategies tailored to their specific contexts. It integrates local knowledge with scientific data to address water scarcity, flooding, and agricultural vulnerability [78, 79].

Traditional ecological knowledge includes hydrology, weather patterns, and land management techniques. For example, indigenous communities often manage ephemeral rivers and grazing patterns based on long-standing practices. Adaptive comanagement bridges local and scientific knowledge through collaborative decision-making [80, 81].

5.7 Gender, vulnerability, and inclusive planning

Gender influences land use and water governance, affecting access, decision-making, and benefits. Women are key users and managers of water and land, yet face systemic barriers to ownership and participation. In Sub-Saharan Africa, women produce up to 80% of food but own less than 15% of agricultural land [82].

Gender-responsive planning incorporates gender analysis throughout policy cycles. It uses participatory tools, gender-disaggregated data, and institutional reforms to ensure inclusion. Inclusive planning also addresses vulnerabilities related to age, disability, and socioeconomic status [83, 84].

Water-related vulnerabilities are magnified by social norms and restricted access to services. Inclusive governance platforms and legal safeguards are essential to build equity and resilience across communities.

The integrated strategies discussed, from zoning and agriculture to governance and inclusivity, form the backbone of a comprehensive approach to water resilience

DOI: http://dx.doi.org/10.5772/intechopen.1012411

in drylands. Blending technical planning with local knowledge, policy reform, and inclusive frameworks ensures that adaptation measures are not only sustainable but also equitable and effective across scales.

6. Conclusion: Forging a resilient future through integrated water-land management in SSA

The escalating challenges of floods, droughts, and fluctuating water tables across Sub-Saharan Africa's (SSA) arid and semi-arid regions demand a profound paradigm shift toward truly integrated land-use strategies. As emphatically demonstrated, these hydro-climatic extremes are no longer isolated events; they are inextricably linked, interacting in complex and often amplifying ways that render conventional, single-hazard responses inadequate. The devastating cycle of post-drought flooding, where parched, impermeable soils transform returning rainfall into destructive flash floods, is a stark testament to this interconnectedness. Such events, increasingly frequent and intense across diverse basins within SSA, highlight how communities face increasingly unpredictable cycles of drought followed by deluge, perpetuating a cycle of loss and vulnerability.

The impacts of these cascading hazards extend far beyond the immediate surface. While episodic floodwaters can offer temporary relief by recharging shallow aquifers, as observed in parts of East Africa, this rapid infiltration also introduces new risks like waterlogging, rising salinity, and infrastructure damage in low-lying or irrigated areas. Conversely, the delayed recovery of many groundwater systems in drylands following prolonged drought, as revealed by GRACE satellite data and corroborated by localized studies, underscores the complex interplay of surface inputs and subsurface inertia. Even after rainfall returns, these vital aquifers can take years to replenish, leaving dependent communities exposed to prolonged water stress. The distinct seasonal fluctuations and long-term trends in terrestrial water storage across various SSA dryland basins further confirm that the region's water reality is shaped by a delicate and dynamic balance between surface inputs and subsurface dynamics, demanding a comprehensive understanding and integrated management.

Despite these critical realities, hazard management across much of SSA regrettably remains fragmented. Flood response often falls under one agency, drought under another, and groundwater governance is frequently neglected entirely. This siloed approach, coupled with fragmented forecasting tools and planning frameworks, severely limits the capacity of nations within the region to anticipate and mitigate these cascading risks, with severe human consequences that range from successive livestock losses and crop destruction to repeated damage to homes. In rural areas near seasonal rivers and depressions, these repeated shocks create a persistent cycle of loss and vulnerability that is difficult to break.

To forge a truly resilient future across SSA, there is an urgent need to adopt a multi-hazard approach that explicitly acknowledges and manages the interactions between water-related hazards. This necessitates

- Integrated policy and governance reforms, i.e., moving beyond fragmented institutional mandates to establish cohesive, overarching frameworks for water and land management that transcend traditional departmental boundaries at national, sub-national, and even transboundary levels. This includes harmonizing policies that govern surface water, groundwater, and land-use planning.

- Adaptive land-use planning by strategically guiding where and how land is developed to fundamentally reduce vulnerability. This is especially vital in SSA's drylands, where climate variability heightens exposure to both floods and droughts. Effective planning includes limiting settlement and critical infrastructure in flood-prone areas, promoting sustainable water management practices, and supporting land uses that enhance drought resilience, such as preserving and restoring vegetation cover, implementing agroforestry, and improving soil organic matter content for enhanced water retention.

- Enhanced early warning systems through the development of sophisticated systems that integrate meteorological forecasts with hydrological and hydrogeological insights, reflecting the intricate connections between surface and subsurface water processes. This will enable better anticipation of cascading events like post-drought flooding and delayed groundwater recovery, allowing for timely, multi-dimensional responses.

- Localized, integrated strategies and community empowerment through fostering participatory approaches that involve local communities in the co-design and implementation of adaptive measures. This includes diverse solutions from improved drainage and water harvesting techniques to drought-resilient cropping calendars and community-managed aquifer recharge initiatives, all designed with the recognition that floods, droughts, and groundwater shifts do not operate in isolation.

- Targeted research and capacity building by investing in robust scientific understanding, systematic data collection (including continuous monitoring of GRACE-derived data, ground-based hydrological observations, and socioeconomic vulnerability assessments), and comprehensive training programs. These efforts are crucial to equip policymakers, practitioners, and local communities with the knowledge, tools, and skills required for integrated risk assessment, adaptive planning, and sustainable resource management.

- Sustainable investment and financing by prioritizing and mobilizing adequate financial resources from national budgets, international development partners, and climate finance mechanisms to support the planning, implementation, and scaling of integrated land-water management projects across SSA.

Ultimately, understanding and proactively managing the dynamic interactions between water-related hazards may prove to be the most decisive factor in building long-term resilience for dryland communities across Sub-Saharan Africa. By embracing a holistic, multi-hazard perspective within their land-use strategies, nations across the continent can move beyond reactive crisis management toward a more secure, sustainable, and prosperous future, better equipped to face the intensifying impacts of climate change.

Author details

Anwarelsadat Elmahal[1*], Wifag Hassan Mahmoud[2], Ahmed Abdalla[3], Etaib Saeed Ganawa[1] and Abdelrahim Salih[4]

1 Faculty of Geographical and Environmental Sciences, University of Khartoum, Sudan

2 Water Harvesting Center, Faculty of Engineering Science, University of Nyala, Nyala, Sudan

3 Department of Civil and Environmental Engineering, Louisiana State University, Baton Rouge, USA

4 Department of Geography, King Faisal University, Al-Ahsa, Saudi Arabia

*Address all correspondence to: anwareltayeb@gmail.com

References

[1] International Union for Conservation of Nature (IUCN). IUCN Annual Report 2012. Gland, Switzerland: IUCN; 2012. [Online]. Available from: https://iucn.org/content/iucn-annual-report-2012 [Accessed: July 12, 2025]

[2] Ouessar M, Gabriels D, Tsunekawa A, Evett S. Water and Land Security in Drylands: Response to Climate Change. Dordrecht, Netherlands: Springer; 2012

[3] Singh RB, Kumar A. Climate variability and water resource scarcity in drylands of Rajasthan, India. Geoenvironmental Disasters. 2015;**2**(1):7. DOI: 10.1186/s40677-015-0018-5

[4] Food and Agriculture Organization of the United Nations (FAO). Drought Characteristics and Management in North Africa and the Near East. Cairo, Egypt: FAO; 2018

[5] Ouessar M, Gabriels D, Tsunekawa A, Evett S. Water and Land Security in Drylands: Response to Climate Change. Cham, Switzerland: Springer; 2017. pp. 1-366

[6] Dapilah F, Nielsen JØ. Climate change extremes and barriers to successful adaptation outcomes: Disentangling a paradox in the semi-arid savanna zone of northern Ghana. Ambio. 2020;**49**(8):1437-1449. DOI: 10.1007/s13280-019-01275-x

[7] Hermans K, McLeman R. Climate change, drought, land degradation and migration: Exploring the linkages. Current Opinion in Environment Sustainability. 2021;**50**:236-244. DOI: 10.1016/j.cosust.2021.04.013

[8] UN News. Flooding from Sudan Dam Collapse Worsens Humanitarian Crisis. UN News; 2024. [Online]. Available from: https://news.un.org/en/story/2024/08/1153651 [Accessed: July 13, 2025]

[9] Coly SM, Zorom M, Leye B, Karambiri H, Guiro A. Learning from history of natural disasters in the Sahel: A comprehensive analysis and lessons for future resilience. Environmental Science and Pollution Research. 2024;**31**(28):40704-40716. DOI: 10.1007/s11356-023-28989-6

[10] World Bank. Chad: Targeted Policy Reforms Can Boost Resilience to Climate Change and Flooding. Washington, DC, USA: World Bank; 2022. [Online]. Available from: https://www.worldbank.org/en/country/chad/publication/chad-targeted-policy-reforms-can-boost-resilience-to-climate-change-and-flooding [Accessed: July 14, 2025]

[11] Tharwat MIH, Modiba MM, Simon B. Effects of climate change on Sudan's water resources. Journal of Central European green innovation. 2023;**11**(3):Art. no. 3. DOI: 10.33038/jcegi.4524

[12] Sutcliffe G, Dugdale G, Milford JR. The Sudan floods of 1988. Hydrological Sciences Journal. 1989;**34**(3):355-364. [Online]. Available from: https://www.tandfonline.com/doi/abs/10.1080/026266689 09491339 [Accessed: July 14, 2025]

[13] Davies HRJ, Walsh RPD. Historical changes in the flood hazard at Khartoum, Sudan: Lessons and warnings for the future. Singapore Journal of Tropical Geography. 1997;**18**(2):123-140. DOI: 10.1111/1467-9493.00012

[14] Williams M, Nottage J. Impact of extreme rainfall in the Central Sudan

during 1999 as a partial analogue for reconstructing early holocene prehistoric environments. Quaternary International. 2006;**150**(1):82-94. DOI: 10.1016/j.quaint.2006.01.009

[15] Mahmood MI, Elagib NA, Horn F, Saad SAG. Lessons learned from Khartoum flash flood impacts: An integrated assessment. Science of the Total Environment. 2017;**601-602**:1031-1045. DOI: 10.1016/j.scitotenv.2017.05.260

[16] Hassan MR, Li F, Zhang Y, Salmi K. A review: Types of floods causes and their impact in Khartoum, Sudan. International Journal of Scientific and Research Publications. 2023;**13**(9):251-260. DOI: 10.29322/ijsrp.13.09.2023.p14129

[17] Babiker IS, Mohamed MAA. Use of NRCS-curve number method for peak discharge estimation in Sharqu ElNeil locality, Khartoum, Sudan. Arabian Journal of Geosciences. 2019;**12**(17):541. DOI: 10.1007/s12517-019-4685-5

[18] Malede DA et al. Understanding flood and drought extremes under a changing climate in the Blue Nile Basin: A review. Environmental and Sustainability Indicators. 2025;**26**:100638. DOI: 10.1016/j.indic.2025.100638

[19] Nile Basin Initiative (NBI). Flood Preparedness and Early Warning (FPEW). Entebbe, Uganda: NBI; 2012. [Online]. Available from: https://nilebasin.org/projects/flood-preparedness-and-early-warning-fpew [Accessed: July 15, 2025]

[20] Belal A-A, El-Ramady HR, Mohamed ES, Saleh AM. Drought risk assessment using remote sensing and GIS techniques. Arabian Journal of Geosciences. 2014;**7**(1):35-53. DOI: 10.1007/s12517-012-0707-2

[21] Siddig K, Stepanyan D, Wiebelt M, Grethe H, Zhu T. Climate change and agriculture in the Sudan: Impact pathways beyond changes in mean rainfall and temperature. Ecological Economics. 2020;**169**:106566. DOI: 10.1016/j.ecolecon.2019.106566

[22] Ahmed SM. Impacts of drought, food security policy and climate change on performance of irrigation schemes in sub-saharan Africa: The case of Sudan. Agricultural Water Management. 2020;**232**:106064. DOI: 10.1016/j.agwat.2020.106064

[23] Rohde MM et al. Groundwater-dependent ecosystem map exposes global dryland protection needs. Nature. 2024;**632**(8023):101-107. DOI: 10.1038/s41586-024-07702-8

[24] Jenkins K, Dobson B, Decker C, Hall JW. An integrated framework for risk-based analysis of economic impacts of drought and water scarcity in England and Wales. Water Resources Research. 2021;**57**(8):e2020WR027715. DOI: 10.1029/2020WR027715

[25] Barbosa HA. Understanding the rapid increase in drought stress and its connections with climate desertification since the early 1990s over the Brazilian semi-arid region. Journal of Arid Environments. 2024;**222**:105142. DOI: 10.1016/j.jaridenv.2024.105142

[26] European Commission, Joint Research Centre (EC-JRC). JRC Analytical Report: Drought and Floods in Sudan – Situation and Outlook (June 2019) – Sudan. Ispra, Italy: EC-JRC; 2019. [Online]. Available from: https://reliefweb.int/report/sudan/jrc-analytical-report-drought-and-floods-sudan-situation-and-outlook-june-2019 [Accessed: July 17, 2025]

[27] Mohmmed A et al. Analysis of drought and vulnerability in the North Darfur region of Sudan. Land

Degradation and Development. 2018;**29**(12):4424-4438. DOI: 10.1002/ldr.3180

[28] Kevane M, Gray L. Darfur: Rainfall and conflict. Environmental Research Letters. 2008;**3**(3):034006. DOI: 10.1088/1748-9326/3/3/034006

[29] United Nations Environment Programme (UNEP). Sudan: Post-Conflict Environmental Assessment. Nairobi, Kenya: UNEP; 2007

[30] Intergovernmental Panel on Climate Change (IPCC). Desertification. In: Climate Change and Land: IPCC Special Report on Climate Change, Desertification, Land Degradation, Sustainable Land Management, Food Security, and Greenhouse Gas Fluxes in Terrestrial Ecosystems. Cambridge: Cambridge University Press; 2022. pp. 249-344. DOI: 10.1017/9781009157988.005

[31] Elmahal A-E, Babikir I, Fadul H. Impacts of Soil Water Rise on Orchard Fields in El-Zaidab Agricultural Scheme. Sudan: River Nile State; 2014

[32] Adam AM, Abdo GM. Water and hydrology (surface and groundwater). In: Salih AM, editor. Water Resources of the Sudan. 1st ed. Khartoum: Water Research Center, University of Khartoum; 2017

[33] Nayl K et al. Hydrogeophysical investigations for groundwater potentiality in arid and semiarid zones: A case study in the Upper River Atbara watershed, eastern Sudan. In: Groundwater Frontiers - Techniques and Challenges. London, UK: IntechOpen; 2024. DOI: 10.5772/intechopen.1007086

[34] Taye MT, Willems P. Identifying sources of temporal variability in hydrological extremes of the upper Blue Nile basin. Journal of Hydrology. 2013;**499**:61-70. DOI: 10.1016/j.jhydrol.2013.06.053

[35] Omer A. Focus on groundwater in Sudan. Environmental Geology. 2002;**41**(8):972-976. DOI: 10.1007/s00254-001-0476-9

[36] Gleeson T, Wada Y, Bierkens MFP, van Beek LPH. Water balance of global aquifers revealed by groundwater footprint. Nature. 2012;**488**(7410):197-200. DOI: 10.1038/nature11295

[37] Abdalla A, Naeimi M, Weigelt M, Müller J. On the Hydrological Mass Variations in Sudan: Based on GRACE Satellite Gravimetry and Global Hydrological Models2018. DOI: 10.13140/RG.2.2.18568.32003

[38] World Bank. Transforming Agriculture in the Sahel: What Would it Take? Washington, DC, USA: World Bank; 2020. [Online]. Available from: https://zcralliance.org/resources/item/transforming-agriculture-in-the-sahel-what-would-it-take/ [Accessed: July 16, 2025]

[39] Morland A. Pastoralism and its Future. The New Humanitarian. 2017. [Online]. Available from: https://www.thenewhumanitarian.org/analysis/2017/10/24/pastoralism-and-its-future [Accessed: July 15, 2025]

[40] Basupi LV, Quinn CH, Dougill AJ. Pastoralism and land tenure transformation in sub-Saharan Africa: Conflicting policies and priorities in Ngamiland, Botswana. Land. 2017;**6**(4):Art. no. 4. DOI: 10.3390/land6040089

[41] SAIS Review of International Affairs. Don't Call it Farmer-Herder Conflict. Washington, DC, USA: Johns Hopkins University; 2020. [Online]. Available from: https://saisreview.sais.jhu.edu/

dont-call-it-farmer-herder-conflict/ [Accessed: July 15, 2025]

[42] Dixon J, Gulliver A, Gibbon D. Farming Systems and Poverty: Improving Farmers' Livelihoods in a Changing World. Rome, Italy, Washington, DC, USA: Food and Agriculture Organization of the United Nations (FAO), The World Bank; 2001. [Online]. Available from: https://www.fao.org/4/y1860e/y1860e04.htm [Accessed: July 15, 2025]

[43] Hirwa H et al. Climate change–drylands–food security nexus in Africa: From the perspective of technical advances, challenges, and opportunities. Frontiers in Environmental Science. 2022;**10**:1-16. DOI: 10.3389/fenvs.2022.851249

[44] Yeleliere E, Antwi-Agyei P, Guodaar L. Farmers response to climate variability and change in rainfed farming systems: Insight from lived experiences of farmers. Heliyon. 2023;**9**(9):1-15. DOI: 10.1016/j.heliyon.2023.e19656

[45] Adeyemi D. Changing weather patterns and its effect on food production in sub-Saharan Africa. Global Warming is Real. 2024. [Online]. Available from: https://globalwarmingisreal.com/2024/12/04/sub-saharan-agriculture-changing-weather-patterns/ [Accessed: July 15, 2025]

[46] Richards S. Conflict in Pastoralist Areas of Sub-Saharan Africa. Medford, MA, USA: Tufts University, Feinstein International Center; 2005. [Online]. Available from: http://fic.tufts.edu/publication-item/conflict-in-pastoralist-areas-of-sub-saharan-africa/ [Accessed: July 15, 2025]

[47] Elouafi I. Drylands under Pressure: Science and Solutions for Global Stability. Geneva, Switzerland: United Nations Office for Disaster Risk Reduction (UNDRR); 2013. [Online]. Available from: https://www.preventionweb.net/news/drylands-under-pressure-science-and-solutions-global-stability [Accessed: July 15, 2025]

[48] Sithole A, Olorunfemi OD. Sustainable agricultural practices in sub-Saharan Africa: A review of adoption trends, impacts, and challenges among smallholder farmers. Sustainability. 2024;**16**(22):Art. no. 22. DOI: 10.3390/su16229766

[49] Tal A. The environmental impacts of overpopulation. Encyclopedia. 2025;**5**(2):Art. no. 2. DOI: 10.3390/encyclopedia5020045

[50] Mbeti S. The Roadblocks to Sustainable Development in Sub-Saharan Africa. Nairobi, Kenya: Africa Sustainability Matters; 2019. [Online]. Available from: https://africasustainabilitymatters.com/the-roadblocks-to-sustainable-development-in-sub-saharan-africa/ [Accessed: July 15, 2025]

[51] Shisanya CA. Editorial: Scaling-up sustainable land management in the drylands of sub-Saharan Africa. Frontiers in Environmental Science. 2024;**12**:1426378. DOI: 10.3389/fenvs.2024.1426378

[52] Global Water Partnership. The Nexus Approach. Stockholm, Sweden: GWP; 2019. [Online]. Available from: https://www.gwp.org/en/GWP-Mediterranean/WE-ACT/Programmes-per-theme/Water-Food-Energy-Nexus/the-nexus-approach-an-introduction/ [Accessed: July 15, 2025]

[53] Thorn JPR et al. Dryland Nature-Based Solutions for Informal Settlement Upgrading Schemes in Africa. Leeds, U.K.: University of Leeds; 2021. [Online]. Available from: https://eprints.whiterose.ac.uk/id/eprint/170735/ [Accessed: July 15, 2025]

[54] Chao O, Li X, Reyes-García V. Exploring the dynamic functions of pastoral traditional knowledge. Ambio. 2025;**54**(6):932-946. DOI: 10.1007/s13280-025-02131-x

[55] Global Network of Civil Society Organisations for Disaster Reduction (GNDR). What Is Risk-Informed Development? London, UK: GNDR; 2021. [Online]. Available from: https://www.gndr.org/resource/risk-informed-development/what-is-risk-informed-development/ [Accessed: July 15, 2025]

[56] Galderisi A, Limongi G. A comprehensive assessment of exposure and vulnerabilities in multi-Hazard urban environments: A key tool for risk-informed planning strategies. Sustainability. 2021;**13**(16):Art. no. 16. DOI: 10.3390/su13169055

[57] Dandoulaki M, Lazoglou M, Pangas N, Serraos K. Disaster risk management and spatial planning: Evidence from the fire-stricken area of Mati, Greece. Sustainability. 2023;**15**(12):Art. no. 12. DOI: 10.3390/su15129776

[58] Sliuzas R, Brussel M, Alam M. Spatial planning and hazard data requirements. In: Geospatial Information for Disaster Risk Reduction. Cham, Switzerland: Springer; 2017. pp. 149-162

[59] Critchley W, Siegert K, Chapman C. Water Harvesting, a Manual Guide for the Design and Construction of Water Harvesting Schemes for Plant Production. FAO: Rome. - References - Scientific Research Publishing. Available from: https://www.scirp.org/reference/ReferencesPapers?ReferenceID=1678343; 1991 [Accessed: July 15, 2025]

[60] Centre for Coordination of Agricultural Research and Development for Southern Africa (CCARDESA). Integrated Soil Fertility Management (ISFM). Gaborone, Botswana: CCARDESA; 2020. [Online]. Available from: https://www.ccardesa.org/integrated-soil-fertility-management-isfm [Accessed: July 15, 2025]

[61] Elmahal AE, Mahmoud WH, Abdalla A, Ibrahim MM. Quantifying the impacts of land use and land cover change on watershed hydrology using spatial cloud computing. Journal of Flood Risk Management. 2025;**18**(1):e70014. DOI: 10.1111/jfr3.70014

[62] Ifejika Speranza C et al. Enhancing the uptake of earth observation products and services in Africa through a multi-level transdisciplinary approach. Surveys in Geophysics. 2023;**44**(1):7-41. DOI: 10.1007/s10712-022-09724-1

[63] Simonovic SP. Bringing future climatic change into water resources management practice today. Water Resources Management. 2017;**31**(10):2933-2950. DOI: 10.1007/s11269-017-1704-8

[64] European Environment Agency (EEA). Floodplain Management: Reducing Flood Risks and Restoring Healthy Ecosystems. Copenhagen, Denmark: EEA; 2020. [Online]. Available from: https://www.eea.europa.eu/highlights/floodplain-management-reducing-flood-risks [Accessed: July 15, 2025]

[65] IUCN. IUCN 2019: International Union for Conservation of Nature Annual Report. 2019. Available from: https://iucn.org/resources/annual-reports/iucn-2019-international-union-conservation-nature-annual-report-2019 [Accessed: July 15, 2025]

[66] Food and Agriculture Organization of the United Nations (FAO). Conservation Agriculture - Revised Version. Rome,

Italy: FAO; 2021. [Online]. Available from: https://openknowledge.fao.org/items/3f75f845-f130-474d-87c9-a1037f7cbad2 [Accessed: July 15, 2025]

[67] Dobhal S et al. Global assessment of production benefits and risk reduction in agroforestry during extreme weather events under climate change scenarios. Frontiers in Forests and Global Change. 2024;**7**:1379741. DOI: 10.3389/ffgc.2024.1379741

[68] Reid R, Jablonski K, Pickering T. Community-Based Rangeland Management in Ethiopia's Pastoral Areas: Trends, Best Practices, and Recommendations for the Future. Nairobi, Kenya: International Livestock Research Institute (ILRI); 2021

[69] Shah T. Taming the anarchy: Groundwater governance in South Asia. Irrigation Science. 2009;**27**(3):311-312. [Online]. Available from: https://link.springer.com/article/10.1007/s00271-009-0194-8 [Accessed: July 15, 2025]

[70] Dillon P, Pavelic P, Page D, Beringen H, Ward J. Managed Aquifer Recharge. Australia: CSIRO Publishing; 2009

[71] Mukherji A, Shah T. Groundwater socio-ecology and governance: A review of institutions and policies in selected countries. Hydrogeology Journal. 2005;**13**(1):328-345. DOI: 10.1007/s10040-005-0434-9

[72] Global Water Partnership (GWP). Legal Framework (A2). Stockholm, Sweden: GWP; 2008. [Online]. Available from: https://www.gwp.org/en/learn/iwrm-toolbox/The-Enabling-Environment/Legal-Framework/ [Accessed: July 15, 2025]

[73] GWP. Integrated Water Resources Management. Stockholm: Global water partnership; 2000

[74] UN. Report of the World Summit on Sustainable Development, Johannesburg, South Africa, 26 August-4 September 2002. New York: UN; 2002. Available from: https://digitallibrary.un.org/record/478154 [Accessed: July 16, 2025]

[75] United Nations Development Programme (UNDP). Mainstreaming Climate Change Adaptation into Development Planning: A Guide for Practitioners. New York, NY, USA: UNDP; 2011. [Online]. Available from: https://www.undp.org/publications/mainstreaming-climate-change-adaptation-development-planning-guide-practitioners [Accessed: July 16, 2025]

[76] United Nations Office for Disaster Risk Reduction (UNDRR). Sendai Framework for Disaster Risk Reduction 2015-2030. Geneva, Switzerland: UNDRR; 2015. [Online]. Available from: https://www.undrr.org/publication/sendai-framework-disaster-risk-reduction-2015-2030 [Accessed: July 16, 2025]

[77] Cook C, Bakker K. Water security: Debating an emerging paradigm. Global Environmental Change. 2012;**22**(1):94-102. DOI: 10.1016/j.gloenvcha.2011.10.011

[78] Forsyth T. Community-Based Adaptation. London, UK, New York, USA: Routledge; 2017

[79] United Nations Development Programme (UNDP). Community-Based Adaptation to Climate Change. New York, NY, USA: UNDP; 2010. [Online]. Available from: https://www.undp.org/publications/community-based-adaptation-climate-change [Accessed: July 16, 2025]

[80] Ayers AL, Kittinger JN, Imperial MT, Vaughan MB. Making the transition

to co-management governance arrangements in Hawai'i: A framework for understanding transaction and transformation costs. International Journal of the Commons. 2017;**11**(1):330-354. DOI: 10.18352/ijc.709

[81] Plummer R, Crona B, Armitage DR, Olsson P, Teng M, Yudina O. Adaptive Comanagement: A systematic review and analysis. Ecology and Society. 2012;**17**(3). Art. No. 11. Available from: https://www.jstor.org/stable/26269066 [Accessed: July 16, 2025]

[82] Meinzen-Dick R, Quisumbing A, Doss C, Theis S. Women's land rights as a pathway to poverty reduction: Framework and review of available evidence. Agricultural Systems. 2019;**172**:72-82. DOI: 10.1016/j.agsy.2017.10.009

[83] UN Women. Turning Promises into Action: Gender Equality in the 2030 Agenda for Sustainable Development. New York, NY, USA: UN Women; 2018. [Online]. Available from: https://www.unwomen.org/en/digital-library/publications/2018/2/gender-equality-in-the-2030-agenda-for-sustainable-development-2018 [Accessed: July 16, 2025]

[84] World Bank. World Development Report 2019: The Changing Nature of Work. Washington, DC, USA: World Bank; 2019. [Online]. Available from: https://www.worldbank.org/en/publication/wdr2019 [Accessed: July 16, 2025]

Chapter 6

Appropriate Mini UAV Technology for Micro-Watershed Land-Use Management Toward a Sustainable Future

Rattana Boonprasert

Abstract

Micro-watersheds, defined as first-order drainage units under 5 km^2, provide a spatially coherent context for localized land-use (LU) and hydrological management. Their smaller scale and homogeneous LU patterns enable more targeted interventions compared to larger watersheds, which often span multiple governance units and present complex hydrological hierarchies. The adoption of UAV-based sensing technologies has significantly enhanced the quality and resolution of spatial inputs required for micro-watershed assessments, particularly for mapping terrain variability, monitoring vegetation stress, and detecting thermal hot spots. Recent studies have shown that UAV-derived datasets, including DEMs, NDVI, soil moisture indices, and land surface temperature (LST), can substantially improve the representation of biophysical processes in hydrological and spatial models. However, to optimize operational efficiency, careful consideration of unmanned aerial vehicles (UAVs) type, payload ratio, and power supply mechanisms is required, particularly in rugged terrain. Alternative and complementary data sources, such as high-resolution satellite imagery, LiDAR, and participatory monitoring, can further improve the accuracy of LU classification and change detection. Advanced modeling approaches, including physics-informed machine learning and graph-based methods, enable the integration of these high-resolution UAV datasets with fundamental hydrological principles, thereby strengthening predictive capacity for environmental planning and climate-resilient LU management. Overall, UAV-based micro-watershed assessment offers an innovative pathway for sustainable land and water resource management, allowing the identification of critical areas such as erosion-prone zones, thermal stress hot spots, and water-limited subcatchments, and supporting evidence-based decision-making at the local scale.

Keywords: micro-watershed, unmanned aerial vehicles (UAVs), land-use assessment, land surface temperature (LST), Structure-from-Motion (SfM)

1. Introduction

1.1 Overview of the application of UAVs in micro-watersheds

1.1.1 Definition and comparative watershed vs. micro-watershed

A watershed, also referred to as a catchment or drainage basin, is a hydrological unit defined by natural topographic boundaries such as ridgelines, encompassing all land that drains into a common outlet, such as a river, lake, or ocean [1]. In contrast, a micro-watershed is a smaller, nested component within a watershed, typically covering less than 5 km^2 [2]. While watersheds may extend from tens to thousands of square kilometers and exhibit complex drainage hierarchies, micro-watersheds are more uniform in scale, allowing for more consistent hydrological responses to precipitation events [3]. Geomorphologically, watersheds consist of varied terrain and heterogeneous hydrological flows across subbasins [2]. Micro-watersheds, however, present simpler physical structures dominated by rill erosion or runoff behavior. Based on spatial scale and stream order, watersheds are further classified into smaller subunits. Mini-watersheds generally cover 1–10 km^2 and are associated with second-order streams, whereas micro-watersheds comprise areas of less than 10 km^2, typically defined by first-order drainage networks. These small-scale hydrological units enable the implementation of precision land and water management practices, particularly in heterogeneous or resource-limited environments, as shown in **Table 1**. From a socioeconomic and land-use (LU) perspective, watersheds may encompass diverse LU types, including urban, rural, and industrial zones spanning multiple governance units and reflecting varied economic activities [4]. Micro-watersheds, on the other hand, are typically associated with homogeneous LU systems such as smallholder agriculture, forested catchments, or rainfed farming communities, often managed under a single jurisdiction. In terms of planning and intervention, the watershed scale supports

Term	Definition	Scale/hydrological criteria/unit	Relevance
Watershed	A topographically delineated land area where all precipitation converges to a common outlet (e.g., a river or lake) through surface and subsurface flows. These function as natural hydrological units, described by their terrain, soil, vegetation, and drainage patterns.	For planning, typically 50–-500 km^2; it may include many subwatersheds.	[1–3, 5, 6]
Mini-watershed	A small catchment unit is often defined as being between 1–10 km^2, representing distinct drainage from first- and second-order streams, and is useful in targeted land and water management.	Approximately 1–10 km^2, second-order streams.	[3, 6]
Micro-watershed	An even smaller hydrological unit (0.1–1 km^2) with unique land–water interactions, suitable for precision interventions in soil conservation or ecohydrology.	Approximately 0.1–1 km^2, first- and second-order streams.	[1–3, 5]

Table 1.
Definitions of watershed, mini-watershed, and micro-watershed [1–3, 5, 6]

integrated ecological and socioeconomic decision-making but poses challenges in terms of modeling complexity and governance coordination. Conversely, micro-watersheds are ideal for high-resolution UAV-based spatial analysis and localized watershed management [5], as shown in **Table 2**. However, their findings may not be easily generalized to broader hydrological systems. Classification frameworks for watersheds often employ hierarchical hydrologic unit codes (HUCs) [6], while micro-watersheds may be prioritized using criteria such as slope gradient, drainage density, land-use intensity, or erosion susceptibility.

In recent years, the integration of UAVs technologies has significantly strengthened the delineation and monitoring of micro-watershed environments. UAV platforms provide high-resolution spatial data, including digital elevation models (DEMs), vegetation indices (VI), and surface soil moisture estimates, thereby enhancing the accuracy of hydrological analyses and erosion assessment at fine spatial scales [7]. Incorporating UAV-derived inputs into watershed modeling frameworks promotes more precise and sustainable LU planning, supporting evidence-based decision-making for ecosystem conservation and climate-resilient resource management. While both watersheds and micro-watersheds are hydrologically defined catchment units, they differ significantly across scale, geomorphology, hydrologic behavior, socioeconomic context, modeling capability, and applicability of management strategies [1]. Watersheds provide a broader context for regional water resource planning, whereas micro-watersheds enable detailed, localized interventions, especially when paired with high-resolution tools such as UAV-based remote sensing and spatial modeling.

The term mini-watershed (often used interchangeably with micro-watershed in various studies) refers to a small-scale hydrological unit that captures water within a confined catchment area. These systems are particularly significant in regions where land fragmentation, diverse terrain, or localized climate variability

Criteria	Watershed	Micro-watershed
Area/scale	Typically, > 10 Km^2.	1–10 Km^2.
Boundary scope	Encompasses large rivers or river systems.	Subset of watershed, focusing on small tributaries [8].
Morphology	Includes varied topography, slope, and features.	Usually homogeneous or semi-homogeneous terrain [9].
Hydrology	Large-scale runoff modeling.	Suitable for small-scale runoff and soil erosion studies [10].
Land-use focus	Regional/national-level policy and planning.	Localized LU, agricultural management [11].
Socioeconomic context	Integrates broader development policies.	Community-level engagement and impact [12].
Classification systems	Based on the catchment hierarchy and river networks.	Based on topography, slope, and drainage lines [8, 9].
Limitations	Complex to model and manage due to scale.	May overlook upstream or external influences [13].
Strengths	Strategic for large-scale conservation and water planning.	Effective for decentralized LU and soil–water management [14, 15].

Table 2.
Criteria/definition and comparative watershed vs. micro-watershed [1–3].

necessitates precise environmental management. Due to their limited spatial scale, mini-watersheds allow for detailed planning of soil and water conservation measures, precision agriculture, and community-level water resource interventions. UAV technologies are increasingly employed to monitor these units because of their ability to generate high-resolution topographic and hydrological data that align with the fine spatial scale of mini-watersheds.

In summary, while both watershed and mini-watershed share the same hydrological foundation, they differ markedly in scale and application. The mini-watershed serves as a granular operational unit that bridges the gap between large-scale hydrological modeling and field-level implementation, thereby offering a critical framework for sustainable and localized natural resource management, as shown in **Tables 1** and **2**.

1.1.2 Conceptual foundation of UAV-based micro-watershed assessment

This chapter establishes a robust theoretical foundation for employing UAVs in micro-watershed land-use analysis. UAV platforms equipped with Structure-from-Motion (SfM) photogrammetry and LiDAR enable the creation of subcentimeter DEMs and high-resolution vegetation mapping capabilities that are critical for resolving fine-scale terrain features, such as gullies and heterogeneous LU patterns, which satellite imagery cannot capture [16, 17]. For instance, UAVs' SfM techniques have demonstrated elevation accuracy comparable to terrestrial LiDAR systems, with median elevation differences of less than 0.1 m, reliably reproducing surface slopes across complex headwater channels [17].

The conceptual foundation of UAV-based micro-watershed assessment in this chapter is as follows:

1. Rationale and scope: UAVs, particularly those using SfM photogrammetry or LiDAR, enable subcentimeter resolution DEMs and highly detailed LU mapping. These capabilities are essential for analyzing micro-watershed terrains, detecting rill and gully erosion, and identifying heterogeneous LU pattern tasks, where traditional remote sensing is often limited [7, 17].

2. Sensor integration for LU insights: UAVs equipped with multispectral, hyperspectral, thermal, RGB, and LiDAR sensors allow multi-layered environmental analysis. SfM-generated DEMs provide terrain data for watershed delineation and erosion risk assessment. Multispectral imagery supports calculation of vegetation indices (e.g., Normalized Difference Vegetation Index NDVI), which are widely used to estimate vegetation cover, density, and overall greenness [15]. In contrast, hyperspectral imagery provides much finer spectral resolution, thereby improving discrimination among different land-cover (LC) types and in some cases allowing differentiation of plant species or vegetation conditions. Combining structural information from Light Detection and Ranging (LiDAR) with hyperspectral data has repeatedly been demonstrated to enhance vegetation or land-cover classification performance, often yielding substantially higher classification accuracies than using either data source alone. For instance, a recent study reported overall classification accuracies exceeding 97 % when fusing LiDAR and hyperspectral data — significantly better than results from spectral data alone [54, 58].

3. Workflow and processing architecture: A typical UAV data pipeline includes the following: (1) image acquisition, (2) DEM construction via SfM or LiDAR, (3)VI and orthomosaics for accurate measurement generation, and (4) spatial analysis in GIS platforms (e.g., ArcGIS, QGIS). This structured approach supports high-precision LU mapping across inaccessible or rugged terrains [7, 21].

4. Integrating multiple sensors, including RGB, multispectral, hyperspectral, thermal, and LiDAR, enables the derivation of VI (e.g., NDVI, NDRE), erosion risk assessments, and orthomosaic generation. Workflow efficiency is further enhanced by GIS-based spatial analysis and machine learning classifiers (e.g., random forest (RF), convolutional neural network (CNN)), which facilitate accurate LU mapping in challenging or inaccessible micro-watershed terrain.

5. Despite their strengths, such as extremely high spatial resolution (often < 5 cm), rapid deployment, and cost-efficiency compared to traditional surveying, UAV-based approaches have notable limitations [7]. Operational challenges include UAV canopy occlusion, high computational demands for processing photogrammetric point clouds, and regulatory or logistical constraints. Nevertheless, when fused with satellite or climatic data, UAV-derived information enhances the scalability and applicability of LU and hydrological models, promoting more informed, sustainable decision-making in agro-ecological and watershed management contexts [2, 19].

6. Rationale and scope: UAVs, particularly those using SfM photogrammetry or LiDAR, enable subcentimeter resolution DEMs and highly detailed LU mapping. These capabilities are essential for analyzing micro-watershed terrains, detecting rill and gully erosion, and identifying heterogeneous LU patterns – tasks where traditional remote sensing is often limited [17].

7. Strengths of UAV-based approaches: UAV offers high spatial resolution (< 5 cm GSD), capable of capturing microtopographic features; cost-effectiveness and flexibility, allowing more frequent data collection than satellite-based platforms; and temporal adaptability, enabling scheduling to align with key crop or hydrological phases [19].

8. Limitations and considerations: High data volume and processing requirements necessitate advanced computing and specialized knowledge. Vegetation occlusion may affect DEM quality in densely vegetated regions unless LiDAR is used. Operational constraints, such as battery limits, weather, and flight regulations, may hinder deployment [19].

9. Conceptual model structure: The proposed model integrates a multi-sensor UAV platform (RGB, multispectral, thermal, LiDAR), DEM, and orthomosaic [22]. Generation is achieved via SfM or LiDAR. Vegetation and erosion indices (EI) include NDVI, NDRE, slope, land cover coefficients, and machine learning classification using RF or CNN models for LU

segmentation. Data integration [23] involves merging UAV outputs with satellite and climatic data for broader landscape inference. This conceptual model supports decision-making for sustainable micro-watershed land management, including zoning, erosion mitigation, and adaptive LU planning [7].

Beyond aeronautical parameters (the conceptual foundation of UAV-based micro-watershed assessment), UAVs classification must also consider sensor payloads and economic feasibility [24]. Platforms suitable for micro-watershed studies often carry a high-resolution visible imaging sensor (RGB color camera) for orthoimage mosaics, multispectral sensors for vegetation index analysis (e.g., VI, NDVI), the Crop Water Stress Index (CWSI), thermal infrared sensors for LST and evapotranspiration assessments, and LiDAR or stereo photogrammetric systems for generating high-precision DEMs [12, 18, 23, 25].

The economic dimension encompasses acquisition costs, maintenance, operational expenses, and the cost-effectiveness of data acquisition. Smaller UAVs offer cost efficiency, portability, and sufficient sensor payload capacity for most micro-watershed monitoring tasks [26]. In contrast, larger UAVs, while offering extended flight endurance and heavier payload support, often entail complex regulatory requirements and increased logistical burdens [24].

2. A suitable alternative to observe in micro-watershed land-use management using UAVs

2.1 Alternative approaches to UAV in micro-watershed LU management

While UAVs offer high-resolution data for micro-watershed monitoring, integrating alternative remote sensing methods, such as high-resolution satellite imagery, airborne LiDAR, participatory monitoring, data fusion techniques, and hydrological modeling, can provide a more comprehensive and practical approach to LU management. These methods, supported by scholarly research, contribute to a holistic understanding of LU changes and their impacts on micro-watersheds. Participatory monitoring involves local communities in data collection, providing ground-truth information that is vital for validating remote sensing data. This approach not only improves the accuracy of LU maps but also fosters community engagement and ownership in watershed management. Studies have highlighted the importance of incorporating local knowledge and participatory methods in environmental monitoring to enhance the effectiveness of management strategies, as well as to improve LU mapping and change detection [27].

Incorporating hydrological models, such as the Surface Energy Balance Algorithm for Land (SEBAL), with remote sensing data enables the estimation of key variables such as evapotranspiration and soil moisture [28]. These models provide insights into the water balance of micro-watersheds, aiding in the assessment of LU impacts on water resources. The application of SEBAL has been demonstrated in various studies to assess water use and quality in watershed areas. While UAVs offer high-resolution spatial data for micro-watershed monitoring, alternative methods can complement or serve as substitutes, especially when UAVs deployment is constrained by budget, regulatory, or technical limitations.

2.1.1 Satellite imagery

Satellite imagery, such as from Landsat or Sentinel missions, provides valuable temporal and spatial coverage for monitoring LU changes in micro-watersheds. These platforms offer multispectral data that can be used to derive indices such as Normalized Difference Vegetation Index (NDVI) and Normalized Difference Water Index (NDWI) [29], which are indicative of vegetation health and water content, respectively. For instance, NDWI can be calculated using Eq. (1) [8], as shown in **Figure 1**.

$$\text{NDWI} = \frac{(\text{Green} + \text{SWIR})}{(\text{Green} - \text{SWIR})} \quad (1)$$

Where:

NDWI (normalized difference water index), "Green" and "SWIR" represent the reflectance in the green and short-wave infrared bands. These indices assist in assessing vegetation cover and water bodies, which are critical for understanding LU dynamics. Other visible-range UAV indices are shown in **Table 3**.

2.1.2 Ground-based monitoring and field surveys

Traditional ground-based methods, including field surveys and soil sampling, remain fundamental for validating remote sensing data and understanding local LU practices [33]. These approaches provide direct measurements of soil properties, vegetation types, and land management practices, offering insights that remote sensing alone may not capture. Integrating field data with remote sensing enhances the accuracy of LU models and ensures that they reflect on-the-ground realities.

2.1.3 Hydrological and soil erosion modeling

Models such as the Soil and Water Assessment Tool (SWAT) and the Conversion of Land Use and its Effects are widely used to simulate land-use (LU) changes and their impacts on hydrology and soil erosion [34]. These models incorporate various environmental variables, such as precipitation, temperature, soil type, and land management practices, to predict changes in LU and their subsequent effects on water quality and quantity, as shown in **Table 4**.

2.1.4 Participatory approaches and citizen science

Engaging local communities through participatory monitoring and citizen science initiatives can provide valuable data on LU changes (LU change) and environmental conditions [68]. These approaches leverage local knowledge and observations, often facilitated through mobile applications or simple data collection tools, to gather information on LU practices, vegetation changes, and other environmental indicators. Such data can complement remote sensing information and improve the spatial and temporal resolution of LU monitoring efforts.

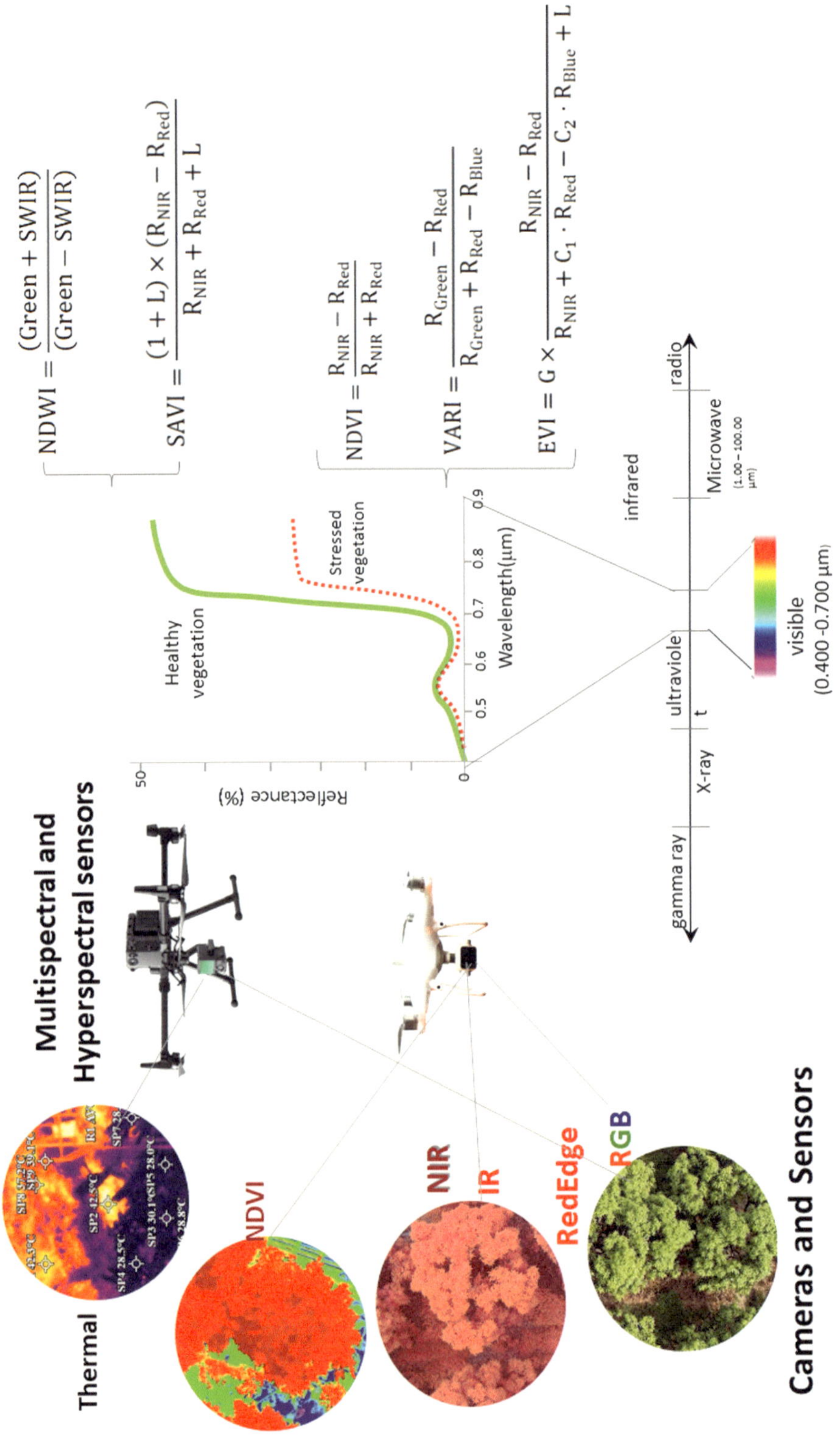

Figure 1.
UAVs offer high-resolution spatial data for micro-watershed monitoring (units under 5 km^2).

Index name	Mathematical formula	UAV remote sensing/relevance
NDVI	$NDVI = \frac{R_{NIR} - R_{Red}}{R_{NIR} + R_{Red}}$	Standard vegetation health metric via multi-spectral UAV [30].
SAVI	$SAVI = \frac{(1+L) \times (R_{NIR} - R_{Red})}{R_{NIR} + R_{Red} + L}$	Reduces soil brightness effects in sparse vegetation scenarios [7, 15].
EVI	$EVI = G \times \frac{R_{NIR} - R_{Red}}{R_{NIR} + C_1 \cdot R_{Red} - C_2 \cdot R_{Blue} + L}$	Improves sensitivity in dense canopies; corrects atmosphere and background [19, 30].
VARI	$VARI = \frac{R_{Green} - R_{Red}}{R_{Green} + R_{Red} - R_{Blue}}$	Useful for RGB-only UAV imagery; resists atmospheric interference [31].
Other visible indices (e.g., S-NDVI, VDVI, TGI).	Various forms combining visible bands.	Enable crop classification and monitoring using standard camera data [32].

Table 3.
Other visible-range UAV indices [7, 15, 19].

Feature	RUSLE	SWAT	UAV-integrated advanced models
Spatial resolution	30–100 m (static grids)	50–500 m (subbasins/ HRUs)	0.05–0.5 m (cm-accurate digital surface model (DSM)/ortho)
Temporal resolution	Annual averages	Daily/monthly	Near-real-time (on-demand flights)
Process physics	Empirical ($A{=}R \cdot K \cdot LS \cdot C \cdot P$)	Semi-process-based (e.g., SCS-CN)	First-principles-driven
Erosion modeling	Sheet/rill erosion only	Limited gully representation	3D mass balance
Data inputs	Lookup tables (C/P factors)	Calibrated parameters (CN, Manning's)	Direct measurement
Uncertainty handling	Sensitivity analysis	GLUE/SUFI-2	Bayesian/EnKF assimilation
Computational demand	Low (Excel implementation)	Moderate (HPC for large basins)	High (GPU edge computing required)
Validation accuracy	NRMSE: 25–40% (outlet sediments)	NSE: 0.5–0.7 (streamflow)	NRMSE < 15% (LiDAR/field validation)
Micro-topography	Cannot resolve features < 10 m.	Limited by DEM resolution	Detects cm-scale gullies/check dams
Best-suited scale	Regional (>100 km^2)	Meso-watersheds (50–500 km^2)	Micro-watersheds (<10 km^2)
Adaptive feedback	None	Manual recalibration	Autonomous UAV resampling

Table 4.
Comparative analysis: traditional vs. UAVs-enhanced models for micro-watershed management [35–38].

2.2 Types/size of UAVs and effective transfer of a sufficient payload in UAVs for micro-watershed land-use management

The principle of effectively transferring a sufficient payload in micro-watershed land-use management depends largely on the structural and operational characteristics of mini and small-scale UAV platforms. Due to their limited physical dimensions, these systems typically possess lower on-board battery capacity, resulting in constrained operational time and payload lift [22]. This chapter has emphasized that the efficiency of UAV-based operations for spatial monitoring and on-site management is strongly influenced by the following: (i) the ratio between payload mass and total take-off mass and (ii) the nature of the on-board power charging type (standard battery, swappable modular battery, solar hybrid, or tethered energy supply) [40]. The effective transfer of a sufficient payload in micro-watershed land-use management: UAVs provide a flexible and high-resolution platform for acquiring spatially explicit data in micro-watershed environments. In this chapter, UAVs are employed to collect multispectral, RGB, and elevation data, enabling detailed characterization of land cover, soil conditions, and hydrological features. These effectively serve as primary inputs for spatial modeling of LU change, allowing the quantification of environmental dynamics [41]:

1) UAVs are categorized based on payload capacity, endurance, and operational range, factors that directly influence their effectiveness in transferring sufficient sensor loads for detailed monitoring. Micro-UAVs (≤2 kg) are deployed for fine-scale surveys with lightweight sensors, Mini-UAVs (2–15 kg) carry high-resolution or multispectral sensors for larger or more complex survey areas, and large UAVs (>15 kg) are reserved for extensive coverage requiring advanced payloads [10]. Appropriate classification of lightweight selection ensures stable flight, optimal data quality, and adequate coverage. Beyond aeronautical parameters, UAV classification must also consider sensor payloads and economic feasibility. Platforms suitable for micro-watershed studies often carry a high-resolution visible imaging sensor (RGB color camera) for orthoimage mosaics, multispectral sensors for vegetation index analysis (e.g., Vegetation Index VI, NDVI, CWSI [42]), thermal infrared sensors for LST and evapotranspiration assessments, and LiDAR or stereo photogrammetric systems for generating high-precision DEMs [23, 26].

The economic dimension encompasses acquisition costs, maintenance, operational expenses, and the cost-effectiveness of data acquisition. Smaller UAVs offer cost efficiency, portability, and sufficient sensor payload capacity for most micro-watershed monitoring tasks [24]. In contrast, larger UAVs, while offering extended flight endurance and heavier payload support, often entail complex regulatory requirements and increased logistical burdens, as shown in **Table 5**.

The effective transfer of a sufficient payload in UAVs for micro-watershed LU management depends largely on the structural and operational characteristics of mini- and small-scale UAV platforms. Due to their limited physical dimensions, these systems typically possess lower on-board battery capacity, resulting in constrained operational time and payload lift. This chapter has emphasized that the efficiency of UAV-based operations for spatial monitoring and on-site management is strongly influenced by the following: (i) the ratio between payload mass and total take-off mass and (ii) the nature of the on-board power charging type (standard battery, swappable modular battery, solar hybrid, or tethered energy supply). As reported by [45], an optimal payload transfer mechanism improves UAV operation by 15–20% in complex terrain when power supply capacity is adjustable in real time. Similarly, [43] demonstrated that

Main features	Types depending on size				
	Nano	Micro	Mini	Small	Large
1. Types depending on size	< 30 mm	< 100 mm	< 300 mm	< 500 mm	> 2 m
2. Altitude (ft AGL)	<200	Up to 200	Up to 3,000	Up to 5,000	Up to 10,000
3. Weight (kg)	<2	<2	2–20	20–150	150–600
4. Payload capacity	<1	1–2.5 kg	1–2.5 kg	1–2.5 kg	>2.5 kg
5. Flight time	<15	15–30 min	30–50 min	25–75 min	>Hours
6. Radius (km)	<5	<25	<50	100–300	>500
7. Spatial coverage in a single flight (km^2)	<0.2	0.2–0.4	0.4-0.8	0.8–3.2	>3.2
8. Endurance (hr)	<1	1–2	1–5	4–15	>24
9. Wind resistivity	<10 m/s	10–15 m/s	8–20 m/s	12 m/s	Highly variable
10. Ability to fly under 11. Windy conditions	Medium	Medium–high	Medium–high	Medium–high	Low

Table 5.
Classification of lightweight unmanned aerial system platforms and their main characteristics [41, 43, 44].

lightweight UAV configurations (<10 kg) can outperform medium-sized platforms in high-elevation micro-watershed landscapes if the payload-to-weight ratio exceeds 0.3 and the energy replenishment time remains below 45 seconds. While implementing micro-watershed management, these technical parameters directly affect the spatial coverage of LU monitoring missions and the frequency of data acquisition for spatial monitoring and on-site management. By incorporating UAV-derived spatial data (e.g., LST, VI, soil moisture) together with environmental drivers (e.g., slope, rainfall, proximity to streams, solar exposure), a spatial mathematical model can be constructed. In this context, the LU change rate within each discrete spatial unit is assumed to be a function of both environmental conditions and UAV operational efficiency. The spatially distributed LU change function is represented in Eq. (2) and Eq. (3) as follows:

$$\frac{\partial L(x,y,t)}{\partial t} = \alpha\left\{\gamma \cdot \frac{P}{M_{\mathrm{UAV}}} + \delta \cdot \frac{C}{T_{\mathrm{chg}}}\right\} \cdot f\left(E(x,y,t)\right) \tag{2}$$

$$f(E) = \beta_1 + LST + \beta_2 + NDVI + \beta_3 + SM + \beta_4 + \cdot SLP + \beta_5 + \cdot R \tag{3}$$

Where:
$L(x, y, t)$ = LU status at the location $(x, y)$and time t; (flight speed or flight time);
P = payload mass (kg);
M_{UAV} = total UAV mass (kg);
C = available power capacity (Wh);
T_{chg} = charging/replacement time (s);
γ *and* δ = UAV efficiency coefficients;
α = a spatial scaling parameter;
$f(E(x, y, t)$ = a function of environmental variables at(x, y, t) such that
LST = LU status at the location $(x, y)$and time t;

NDVI = vegetation index;
SM = soil moisture;
SLP = slope;
R = rainfall; and β's are empirically derived parameters.

This spatial differential equation can be solved numerically to simulate LU evolution across the micro-watershed domain and allows scenarios reflecting different UAV operational payloads and power-charging strategies.

2) UAV size, payload, and battery capacity: Mini- and small UAVs, often defined as weighing between approximately 0.25 kg to 25 kg and having wingspans between 50 cm and 2 m, are valued for their portability and affordability in environmental monitoring but are inherently constrained by limited onboard battery capacity and payload capability [29]. As UAVs increase battery capacity to extend flight time, overall mass increases, which in turn raises energy consumption, creating a trade-off between payload duration and operational efficiency [24, 40]. Moreover, payload capacity is influenced not only by frame size and motor power but also by the number of propellers and structural design [29, 40].

3) Charging and powering strategies: Flight time capability; controlled crash, or parachute. Given battery limitations, research has explored alternative approaches to sustain UAV operations. Battery swapping (hot-swapping): landing and rapidly replacing depleted batteries, minimizing downtime, though it incurs logistical complexity [29]. Wired or wireless recharging at base stations: Inductive or magnetic resonance coupling enables contactless ground recharging, with experimental efficiencies up to ~ 90% at short distances (~10 cm) [29]. Laser-based in-flight recharging (optical wireless power transfer): using directed laser beams or distributed laser charging to power or recharge UAVs mid-air or at perches [29]. Hybrid power systems: combining batteries with solar cells, fuel cells, supercapacitors, or other sources to extend endurance and dynamically manage power [48]. These systems aim to mitigate the limitations of battery energy density, reduce flight interruptions, and improve mission flexibility and duration, especially critical for repeated flights over micro-watersheds for LU mapping or monitoring.

3. Modeling approaches integrated with UAV technology for micro-watershed management

Recent developments in environmental modeling have increasingly incorporated high-resolution datasets obtained from UAV platforms into advanced computational frameworks [46]. These new-generation frameworks facilitate dynamic and spatially explicit assessment of land and water interactions at micro-watershed scales, enabling more precise representation of topographic and biophysical processes. UAVs play an essential role in this context by supplying fine-scale inputs, including DEMs, multispectral VI (e.g., NDVI), and soil moisture estimates, which enhance the accuracy and physical realism of hydrological simulations [47].

3.1 Appropriate data from UAV to support quantitative, model-based planning and assessment

3.1.1 Terrain analysis

1) Slope calculation (from UAVs' DSMs) in Eq. (4):

$$\left(S = \arctan\left(\sqrt{\left(\frac{\partial z}{\partial x}\right)^2 + \left(\frac{\partial z}{\partial y}\right)^2} \right) \right) \tag{4}$$

Where:

Symbol	Name	Definition	UAV data source
S	Slope	Angle between the terrain surface and the horizontal plane (critical for erosion/runoff).	Output of the equation.
z	Elevation	Height above a reference datum (e.g., sea level).	DSMs
$\left(\frac{\partial z}{\partial x}\right)^2$	East–west gradient	Rate of elevation change along the x-axis (longitudinal direction).	Calculated from DSM using finite differences.
$\left(\frac{\partial z}{\partial y}\right)^2$	North–south gradient	Rate of elevation change along the y-axis (latitudinal direction).	Calculated from DSM using finite differences.

Calculation process:

(1) DSMs preprocessing:

- UAV photogrammetry (e.g., SfM) generates a DSM with 1–10 cm resolution.
- Noisy data are filtered using Gaussian smoothing or median filters.

(2) Partial derivatives

- Computed using finite difference methods in Eq. (5) and Eq. (6):

$$\frac{\partial z}{\partial x} \approx \frac{z(x + \Delta x, y) - z(x - \Delta x, y)}{2\Delta x} \tag{5}$$

$$\frac{\partial z}{\partial y} \approx \frac{z(x, y + \Delta y) - z(x, y - \Delta y)}{2\Delta y} \tag{6}$$

$-(\Delta x, \Delta y)$ = DSM grid spacing (e.g., 0.05 m for UAV data).

(3) Slope magnitude ($\|\nabla z\|$ $\|\nabla z\|$) in Eq. (7):

$$||\nabla z|| = \sqrt{\left(\frac{\partial z}{\partial x}\right)^2 + \left(\frac{\partial z}{\partial y}\right)^2} \tag{7}$$

Represents the steepest gradient at each pixel.

(4) Slope angle (S) in Eq. (8):

$$S = \arctan(||\nabla z||) \tag{8}$$

Converts gradient to degrees (°) for practical use (e.g., 15° = high erosion risk).

This approach leverages UAV precision to quantify slope-driven processes in micro-watersheds, enabling targeted conservation strategies that are impossible with satellite/low-resolution DEMs [38], as shown in **Table 6**.

3.2 Appropriate UAVs technology for spatial mathematical hydrological modeling in micro-watersheds

UAVs provide highly detailed spatial inputs that greatly improve the precision of hydrological models in micro-watershed environments. UAV-based remote sensing allows for the rapid acquisition of fine-resolution DEMs, land cover classifications, and surface roughness indicators, which are essential for parameterizing spatially distributed hydrological models. Jamwal and Thakur [4] highlighted that the spatial accuracy of UAV-derived DEMs facilitates improved model-based planning and environmental assessment in wetland and watershed contexts. Similarly, Mazzoleni et al. [9] demonstrated that UAV datasets yielded more reliable hydraulic model outputs compared with coarser elevation data, thereby strengthening flood risk assessments and runoff predictions in small-scale catchments.

In micro-watershed modeling contexts, spatial mathematical formulations are commonly applied by discretizing the terrain into raster cells and explicitly

Aspect	**Traditional DEMs (e.g., SRTM 30 m)**	**UAVs/DSMs (5 cm)**
Accuracy	±5–10 m vertical error.	±1–5 cm vertical error
Spatial resolution	30–90 m pixels.	0.05–0.2 m pixels
Micro-topography	Misses gullies, terraces.	Captures rills < 20 cm wide
Slope error	25–40% in steep terrain.	<5% (validated by LiDAR)
Erosion risk mapping	Slopes > 15° are classified as high risk (requiring contour terracing).	0.05–0.2 m pixels
Drainage planning	Flow direction algorithms use slope to model runoff paths.	Captures rills < 20 cm wide
Infrastructure design	Roads/trails avoid slopes > 30° (UAV detects unstable zones).	±1–5 cm vertical error

Table 6.
Practical applications of UAV terrain analysis data in micro-watershed management [38].

incorporating spatial variables derived from UAV data. Elevation is modeled as a terrain function in Eq. (9):

$$\text{For each cell}(i,j):\begin{cases} z_{i,j} = h(x_i, y_j) \\ S_{i,j} = \sqrt{\left(\frac{\partial h}{\partial x}\right)^2 + \left(\frac{\partial h}{\partial y}\right)^2} \\ n_{i,j} = f(LULC_{i,j}) \\ v_{i,j} = \frac{1}{n_{ij}} H_{i,j}^{2/3} \cdot S_{i,j}^{1/2} \\ \frac{\partial H_{i,j}}{\partial t} = P_{i,j} - \nabla \cdot (v_{i,j} H_{i,j}) - E_{i,j} + I_{i,j} \end{cases} \tag{9}$$

Where:

$P_{i,j}= P \bullet C_{LU\,LC_{i,j}}$ reflects the effect of land use on rainfall–runoff response. This structure enables model-based planning and assessment by linking hydrological processes to spatial patterns of terrain and LU. UAV-derived inputs, therefore, provide a reliable basis for simulating LU change impacts and evaluating alternative management scenarios at the micro-watershed scale. Effective rainfall at cell *(i,j)* is calculated from total rainfall *P*

and runoff coefficient $C_{LU\,LC_{i,j}}$

x_i, y_j = Spatial coordinates of the center of each raster cell in the micro-watershed grid system.

$z_{i,j}$ = Elevation value at cell (i, j), derived from the UAV-based DEM.

$h(x_i, y_j)$ = Terrain elevation function that maps each spatial location to its corresponding height above mean sea level.

$S_{i,j}$ = Local surface slope at each cell, calculated as the spatial gradient of the DEM (i.e., the rate of elevation change in the x and y directions).

$LULC_{i,j}$ LU/LC class of the cell (i, j), derived from UAV orthophotos using image classification $n_{i,j} = f(LULC_{i,j})$

$n_{i,j}$ Spatially varying Manning's roughness coefficient. It depends on the LU type in each cell, through a function $n_{i,j} = f(LULC_{i,j})$

$H_{i,j}$ Water depth in cell (i, j), time t. It represents the hydrological state of the cell and varies continuously over time.

$\frac{\partial H_{i,j}}{\partial t}$ Temporal rate of change in water depth, that is, how fast the depth increases or decreases over time.

P External rainfall amount (mm/h), applied uniformly over the watershed or spatially distributed, where data are available.

$C_{LU\,LC_{i,j}}$ Rainfall–runoff coefficient that adjusts rainfall input according to land-use type in each cell (e.g., agricultural areas vs. forest).

$E_{i,j}$ Evaporation or infiltration loss at cell (i, j), representing water that does not contribute to surface runoff.

$v_{i,j}$ Flow velocity of water in cell (i, j), estimated using the spatial form of Manning's equation.

$Q = \nabla \cdot (v_{i,j} H_{i,j})$ Divergence of flow (i.e., outflow of water from cell (i, j), to adjacent cells).

Inflow to cell (i, j), from upstream neighboring cells, based on local flow direction and velocity fields.

3.3 Methodology: UAVs-based spatial hydrological modeling in micro-watersheds

To simulate hydrological processes in micro-watersheds, high-resolution spatial data are collected using UAVs. UAV-derived DEMs provide accurate terrain information, while orthophotos enable classification of land use/land cover (LULC). Surface roughness metrics are extracted from these datasets to parameterize spatially varying hydraulic resistance, such as Manning's coefficient. The micro-watershed is discretized into raster cells (i, j), each associated with spatial coordinates. The elevation of each grid cell in a digital elevation model (DEM) can be expressed as a surface function z=f(x,y), representing the underlying terrain. Local slope characteristics are derived by computing the partial derivatives of this function in the two horizontal directions, ∂z/∂x and ∂z/∂y These derivatives form the gradient vector, whose magnitude is commonly used to quantify the overall slope steepness of the terrain surface. In practical terrain analysis, the derivatives are typically approximated using finite-difference operators or neighborhood-based surface fitting methods. Such gradient-based computations constitute a fundamental component of digital terrain analysis and hydrological modelling using DEMs.

Hydrological state variables include the water depth and its temporal rate of change, which are determined by the balance of effective rainfall, surface inflow, outflow, and losses due to evaporation or infiltration. Effective rainfall is computed by adjusting measured rainfall *P* according to land use through a runoff coefficient. Flow velocity in each cell is estimated using the spatially distributed Manning's equation in Eq. (10):

$$v_{i,j} = \frac{1}{n_{i,j}} H_{i,j}^{2/3} \cdot S_{i,j}^{1/2} \tag{10}$$

Where the Manning's coefficient is a function of LULC in each cell. Outflow from each cell is expressed as $Q =$, representing spatial redistribution of water across the micro-watershed. Inflow accounts for contributions from upstream neighboring cells, ensuring mass conservation. By integrating UAV-derived elevation, land cover, and roughness information into these spatially explicit equations, the model simulates hydrological processes, captures the influence of terrain and land use on water movement, and enables scenario-based planning and assessment of micro-watershed management practices.

3.4 Physics-informed machine learning (PIML)

Among the most promising approaches is PIML, which embeds physical constraints (such as water or energy balance equations) within machine learning algorithms [48]. When applied to UAV-acquired environmental datasets, PIML enables the generation of model outputs (e.g., runoff or LU suitability) that are not only data-driven but remain consistent with fundamental hydrological principles. Similarly, graph neural network (GNN) frameworks have been increasingly adopted in rainfall–runoff modeling and watershed analysis. In this approach, UAV-derived variables such as terrain slope, vegetation health, and surface drainage patterns are used to define the node attributes of a spatial graph, while the graph structure captures hydrological connectivity among terrain units. Through the integration of UAV-generated datasets, these advanced methods support highly accurate prediction and spatially explicit analysis of landscape responses, thereby facilitating adaptive LU planning, erosion mitigation, and climate-resilient micro-watershed management [48, 49], which incorporates physical laws (e.g., water balance) into machine learning models. When applied to UAV-acquired environmental data, the PIML framework optimizes model accuracy while enforcing hydrological constraints in Eq. (11):

$$h_i^{(k+1)} = \sigma\left(\sum_{j\in N(i)} \frac{1}{\sqrt{d_i d_j}} W^{(k)} h_j^{(k)} + b^{(k)}\right) \tag{11}$$

Where:
$h_i^{(1)}$ = includes UAV-derived variables such as slope, land cover type, and water accumulation zones;
$N(i)$ = represents neighboring zones within the micro-watershed.
The final graph output estimates localized runoff or land degradation risks.

X may include UAV-derived data such as terrain slope (DEMs), vegetation NDVI, and soil moisture. This allows the model to generate runoff or land-use suitability outputs that are both data-driven and physically grounded, which is particularly useful in complex, sensitive micro-watersheds. By leveraging UAV-generated spatial datasets, these advanced models offer unprecedented precision in analyzing and forecasting environmental responses. Their integration supports adaptive LU planning, erosion control strategies, and climate-resilient watershed development, which is especially vital for regions where data scarcity and complex terrain have historically impeded sustainable resource management [47]. This enables localized prediction of runoff, erosion potential, or LU transition risks based on neighboring spatial conditions. UAVs enhance the accuracy of these models by providing high-fidelity data on slope, vegetation, and drainage networks that are otherwise unobtainable in heterogeneous or remote regions [1]. The synergy between UAV technology and these next-generation models forms the foundation for a scalable, data-driven, and environmentally responsive micro-watershed management strategy. This integration allows policymakers, researchers, and land managers to make informed decisions that align with long-term sustainability and climate adaptation goals [Filho F].

4. Appropriate UAV technology for micro-watershed land-use (LU) assessment

4.1 Conceptual foundation of appropriate Mini UAVs technology for micro-watershed LU assessment

UAVs integrated with advanced sensors such as SfM photogrammetry and LiDAR offer unprecedented spatial resolution, enabling subcentimeter DEMs and detailed vegetation characterization [50, 51]. These capabilities are critical for micro-watershed assessments, particularly for capturing terrain variability, gully formation, and fine-scale LU patterns [19, 27] that traditional satellite-based remote sensing fails to detect. UAVs equipped with RGB, multispectral, hyperspectral, thermal, and LiDAR sensors enable the generation of high-resolution orthomosaics,

VI (e.g., NDVI, NDRE), and topographic variables, allowing for spatially explicit analyses of erosion risks, flow direction, and LU classification through GIS platforms.

A proposed conceptual model integrates this sensor suite to produce DEMs and orthomosaics, followed by spatial analysis using machine learning algorithms (e.g., RF, CNN) for LU classification. LiDAR and hyperspectral fusion, in particular, improves classification accuracy by over 30% compared to single-sensor approaches, making it suitable for ecological mapping at the species level [19]. Despite some limitations, such as the need for technical expertise, vegetation occlusion in photogrammetry, and operational constraints, UAV systems present a cost-effective and temporally flexible solution. When combined with satellite or climate datasets, this UAV-based framework can be scaled up for broader landscape monitoring and applied in agro-ecological conservation planning.

Despite their strengths, such as extremely high spatial resolution (often < 5 cm), rapid deployment, and cost-efficiency compared to traditional surveying, UAV-based approaches have notable limitations. Operational challenges include dense canopy occlusion, high computational demands for processing photogrammetric point clouds, and regulatory or logistical constraints. Nevertheless, when fused with satellite or climatic data, UAV-derived information enhances the scalability and applicability of LU and hydrological models, promoting more informed, sustainable decision-making in agro-ecological and watershed management contexts [19, 51] as shown in **Table** 7.

4.2 UAV technology for micro-watershed land-use management toward a sustainable future

4.2.1 UAV-based LST data support the assessment of environmental dynamics

UAV-based LST and its relationship with NDVI covers land use, strengths, limitations, and classifications: 1) definition and land-use applications – LST derived from UAV-mounted thermal sensors quantifies the Earth's radiative surface temperature at ultra-high spatial resolution (0.1–1 m). When integrated with NDVI from multispectral UAV sensors, it enables precise analysis of LU patterns. Key applications include the following: (i) precision agriculture, where LST–NDVI correlations detect crop water stress for optimized irrigation [60]; (ii) urban heat mitigation (UHIs), identifying micro-scale heat islands to guide green infrastructure planning [31]; and (iii) ecosystem monitoring, tracking drought impacts in forests via LST–NDVI anomalies. 2) Strengths, limitations, and classifications: LST derived from UAV-mounted thermal sensors quantifies the Earth's radiative surface temperature at ultra-high spatial resolution (0.1–1 m). When integrated with the NDVI from multispectral UAV sensors, it enables precise analysis of LU patterns. Key applications include: (i) precision agriculture, where LST–NDVI correlations detect crop water stress for optimized irrigation [60]; (ii) UHIs, identifying micro-scale heat islands to guide green infrastructure planning [31]; and (iii) ecosystem monitoring, tracking drought impacts in forests via LST–NDVI anomalies [61], as shown in **Table** 8.

Appropriate UAV technology	Components	Descriptions/outputs	Representative references
UAV platform and sensor selection	UAV types	Multi-rotor, fixed-wing, or hybrid systems are selected based on terrain, area, and operational objectives.	[24, 52, 53]
	Sensors	RGB cameras – orthophotos and 3D terrain via SfM photogrammetry.	[18, 19, 25, 30, 43, 50, 51, 54, 55]
		Multispectral sensors – VI (NDVI, NDRE, NDWI).	
		Thermal cameras: LST, crop water stress, evapotranspiration.	
		LiDAR/radar – canopy structure and elevation (forested watersheds).	
		Microwave radiometers – soil moisture estimation.	
Data Acquisition and Pre-processing	Flight planning	Based on topography, weather, battery endurance, and ground sampling distance (GSD) requirements.	[19, 44]
	Products	DEMs and orthomosaics.	[22, 25 26, 36, 53]
		Vegetation index maps (e.g., NDVI).	
		Thermal maps (surface temperature/water stress).	
		Soil moisture maps (passive radiometry/fusion).	
	Quality control	Use of GCPs, RTK-GNSS, and ground-based validation.	[26, 44, 49, 56, 26]
Data Analysis and LU Modeling	Machine-learning classification	RF and CNNs for LU and vegetation mapping.	[17, 20, 22, 49, 57]
	Terrain/ Hydrological modeling	Flow accumulation, slope thresholds, and erosion zones from UAV DEMs.	[5, 6, 31]
	Water and soil dynamics	Evapotranspiration (thermal imagery), soil moisture for drought/irrigation appraisal.	[17, 20, 40, 57]
Integration and Decision Support	Decision support tools	UAV outputs, combined with GIS layers, satellite data (e.g., Sentinel-2), and existing hydro-ecological models.	[2, 5, 6, 29, 59]
		Multi-scalar visualization (e.g., dashboards); participatory planning and conservation zoning.	[28, 39, 47]
	End users	Local communities, watershed managers, and national agencies are informed for LU optimization and conservation.	[5, 6, 14, 22, 59]

Table 7.
Appropriate UAV technology for micro-watershed LU assessment.

4.2.2 UAV thermal sensing enables high-resolution acquisition of LST imagery, providing detailed spatiotemporal information

That is indispensable for evaluating micro-watershed conditions [60, 62]. When integrated with spatial–mathematical models, UAV-based LST data support the assessment of environmental dynamics, including topography, vegetation cover, soil moisture distribution, and small water bodies (<100 km^2). The fine spatial resolution of UAV thermal sensors facilitates the detection of LU change processes and erosion

Aspect	Description	References
Definition	• LST: Radiative surface temperature measured via UAV thermal sensors (0.1–1 m resolution). • LST refers to the radiative skin temperature of the land's surface as measured by thermal sensing instruments, and it differs from air temperature since it directly relates to the energy exchange at the Earth's surface. LST integrates the effects of surface materials, moisture, vegetation, and urban geometry.	[7, 60]
LU applications	• NDVI-LST synergy: Combines thermal and multispectral data for land surface analysis. • Agriculture: Detecting crop water stress for precision irrigation. • Urban: Micro-scale heat island (UHI) mapping for green infrastructure planning. • Forestry: Drought impact assessment via LST–NDVI anomalies.	[30, 31, 42, 56]
Strengths	• Ultra-high spatial resolution (cm-scale). • Flexible deployment (cloud-adaptive). • Enables evapotranspiration (ET) and soil moisture modeling.	[7, 31, 62]
Limitations	• Limited spatial coverage (unsuitable for large areas). • Atmospheric interference requires ground calibration.	[60, 63]
Classification systems	• NDVI saturation in dense vegetation (use EVI/OSAVI alternatives).	[60]
	• FAO land cover classification (LCCS): Stratifies urban/agricultural zones. • Urban Thermal Field Variance Index: Quantifies UHI intensity.	[31, 60, 62]

Table 8.
UAV-based LST and NDVI: applications, limitations, and classifications.

features such as rill and gully formation, and provides an empirical basis for the quantification of crop water stress and irrigated zones [31]. Consequently, the evaluation of LST images obtained from UAV platforms contributes significantly to evidence-based decision-making for sustainable LU planning and ecosystem conservation at the micro-watershed scale [15].

To operationalize this process, a spatial–mathematical model can be formulated in the form of a multivariate spatial function describing the relationship between LU types and environmental variables derived from UAV data. Let $LST(x, y)LST(x, y)$ $LST(x, y)$ denote the LST at coordinates (x, y)(x, y)(x, y) within the micro-watershed domain $D \subseteq R^2 D$ that defines the spatial relationship in Eq. (12), as shown in **Figure 2**.

$$LST(x,y) = f\Big(LU(x,y),\ SM(x,y),\ NDVI(x,y),\ Elev(x,y),\ Dist_w(x,y)\Big) \tag{12}$$

Where:
$LU(x, y)$ = land-use category
$SM(x, y)$ = soil moisture
$NDVI(x, y)$ = vegetation index computed from UAV multispectral imagery
$Elev(x, y)$ = Elevation
$Dist_w(x, y)$ = distance to nearest water body, and $\varepsilon(x, y)$
$\varepsilon(x, y)$ = spatially autocorrelated error term
$LU(x, y, t)$ = LU type at location (x, y) and time t

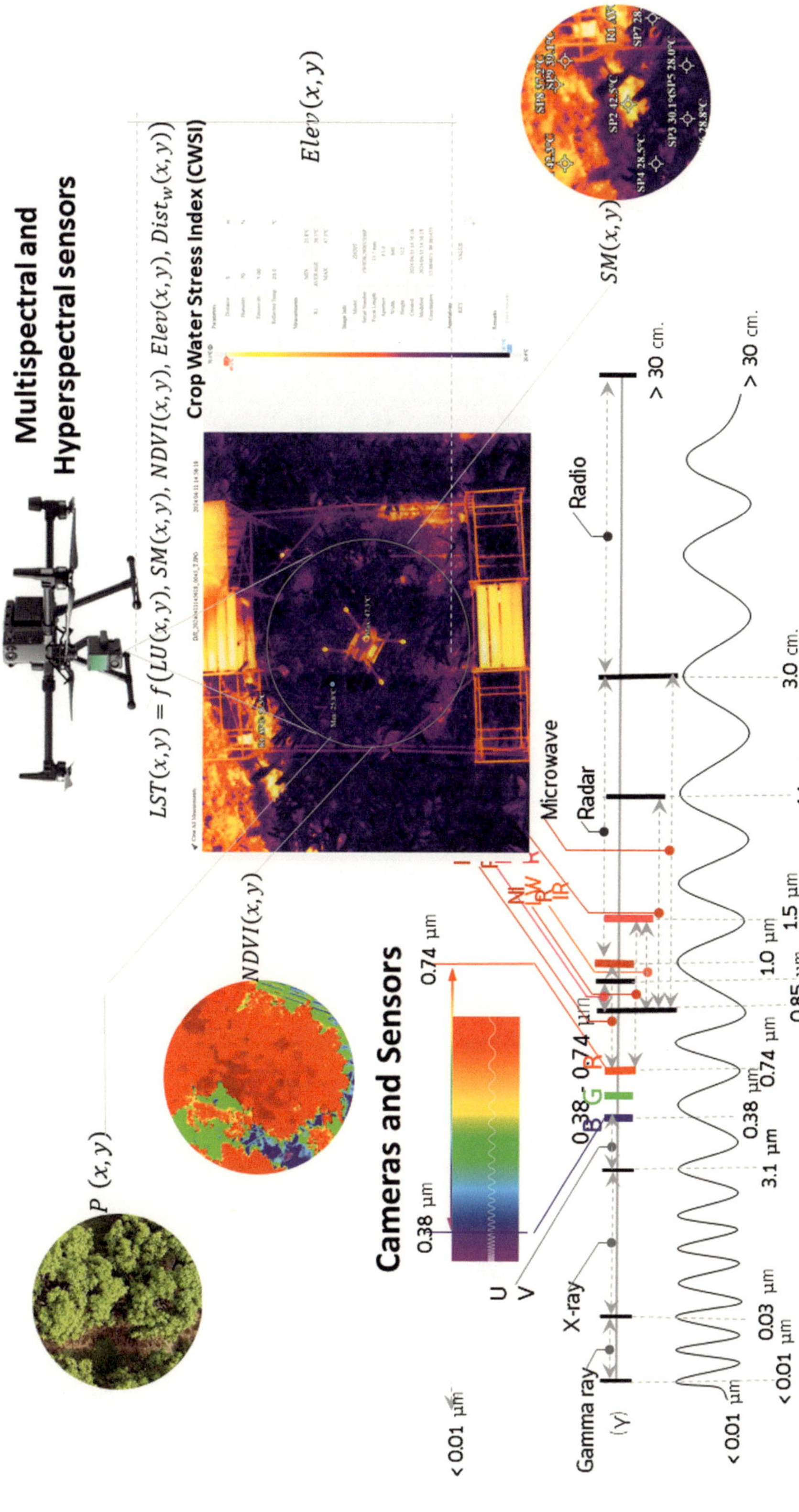

Figure 2.
UAV thermal sensing enables high-resolution acquisition of LST imagery, providing detailed spatiotemporal information.

LST(x, y, t) = land surface temperature
SM(x, y, t) = vegetation index computed from UAV's multispectral imagery
NDVI(x, y, t)= NDVI
Elev(x, y) = elevation
$Dist_w(x, y)$ = distance to the nearest water body
P(x, y, t) anthropogenic pressure (e.g., agricultural intensity)

The function (·) can be parameterized using spatial regression or geostatistical approaches (e.g., geographically weighted regression) to capture the heterogeneity of micro-watershed conditions. Spatiotemporal changes in land use can then be modeled in Eq. (13):

$$\frac{\partial LU(x,y,t)}{\partial t} = g\Big(LST(x,y,t),\ SM(x,y,t),\ \nabla Elev(x,y),\ P(x,y,t)\Big), \tag{13}$$

where:

P(x, y, t) denotes anthropogenic pressure (e.g., agricultural intensity), and g(·) represents the rate of LU transition conditioned by environmental state variables. These formulations allow for the evaluation of LST variability and its influence on LU dynamics under distinct environmental scenarios, thereby providing a robust analytical framework for micro-watershed assessment based on UAV-derived spatial data. Multivariate spatial mathematical system: Let $D \subseteq R^2$ denote the micro-watershed domain.

The primary objective is to describe the dynamics of LU and environmental interactions using UAV-derived spatial variables. System of multivariate equations in Eq. (14), Eq. (15), Eq. (16), and Eq. (17):

$$\begin{aligned} LST(x,y,t) = {} & \alpha_0 + \alpha_1 LU(x,y,t) + \alpha_2 SM(x,y,t) + \alpha_3 NDVI(x,y,t) + \alpha_4 Elev(x,y) \\ & + \alpha_5 Dist_w(x,y) + \varepsilon_1(x,y,t) \end{aligned} \tag{14}$$

$$\begin{aligned} SM(x,y,t) = {} & \beta_0 + \beta_1 LU(x,y,t) + \beta_2 Precip(x,y,t) + \beta_3 Irr(x,y,t) + \beta_4 Elev(x,y) \\ & + \varepsilon_2(x,y,t) \end{aligned} \tag{15}$$

$$NDVI(x,y,t) = \gamma_0 + \gamma_1 LST(x,y,t) + \gamma_2 SM(x,y,t) + \gamma_3 LU(x,y,t) + \varepsilon_3(x,y,t) \tag{16}$$

$$\begin{aligned} \frac{\partial LU(x,y,t)}{\partial t} = {} & \delta_0 + \delta_1 LST(x,y,t) + \delta_2 SM(x,y,t) + \delta_3 NDVI(x,y,t) + \delta_4 \nabla Elev(x,y) \\ & + \delta_5 P(x,y,t) + \varepsilon_4(x,y,t) \end{aligned} \tag{17}$$

5. Spatial model development based on UAV data for micro-watershed management

The spatial model for micro-watershed management using UAV-derived data is developed through three integrated stages: data acquisition and preprocessing, spatial variable formulation, and model implementation and assessment. 1) Data acquisition and preprocessing: High-resolution datasets are acquired using multispectral and thermal UAV sensors. These datasets include LST, NDVI, soil moisture, DSMs, and orthophotos [4]. The images are geometrically corrected and coregistered to a common coordinate system (e.g., UTM) to enable precise overlay and pixel-level comparison. 2) Spatial variable formulation: Based on the UAV's products, spatial variables are constructed to represent environmental conditions at each location (x, y) within the micro-watershed. The following spatial functions are defined.

5.1 Conceptual framework for micro-watershed LU management toward a sustainable future

The proposed conceptual framework for UAV-based micro-watershed assessment integrates high-resolution spatial data with environmental and LU dynamics to support sustainable management decisions. At the foundation, UAV-derived spatial data provide fine-scale measurements of critical variables, including LST, NDVI, soil moisture, topography (elevation), and distance to water bodies [50]. These datasets serve as empirical inputs for modeling the micro-environmental conditions within watersheds smaller than 100 km^2. The second module, environmental variables, processes UAV-derived data to characterize local micro-climatic and ecological conditions. LST, soil moisture, and VI are used to quantify micro-climatic variability, soil health, and vegetation vigor, while elevation and proximity to water bodies provide topographical and hydrological context. These variables collectively influence biophysical processes such as evapotranspiration, runoff, and erosion, forming the basis for predictive modeling of LU dynamics. In the LU dynamics module, environmental variables interact with anthropogenic factors to determine spatial and temporal changes in LU types [7]. Cropland, forest, and grassland distribution, irrigation patterns, crop water stress, and erosion features are modeled as functions of environmental states and human pressures. The system is expressed mathematically through multivariate equations linking LST, NDVI, soil moisture, and topography to LU transitions over time [30]. This integration captures the complex feedback between natural and human-modified processes at the micro-watershed scale. Finally, the micro-watershed assessment and decision-making module synthesize outputs from environmental monitoring and LU modeling to guide sustainable planning and conservation strategies. By explicitly linking UAV-derived measurements, environmental conditions, and LU dynamics, the framework facilitates evidence-based interventions aimed at mitigating soil erosion, optimizing irrigation, and maintaining ecosystem services. Feedback loops from decision-making to LU and environmental modules support adaptive management, enabling continuous monitoring and responsive adjustments to maintain watershed sustainability [64].

5.2 Thermal hot spots

UAV-based LST–NDVI modeling offers a highly effective approach for detecting thermal hot spots at very fine spatial scales. By integrating thermal infrared sensors with multispectral cameras, UAVs allow simultaneous acquisition of LST and NDVI, which improves emissivity estimation and increases the accuracy of temperature calculations [19]. This capability is particularly useful for precision agriculture and land management in micro-scale environments, where detailed identification of surface heterogeneity (e.g., vegetation, bare soil, or impervious surfaces) is required to evaluate local thermal dynamics. In addition, UAV thermal imaging has been successfully applied to assess micro-urban heat islands and residential micro-scale variations [34].

For instance, recent research in urban areas has demonstrated that UAVs can differentiate between high-temperature zones, such as dark rooftops or compacted surfaces, and vegetated areas with significantly lower temperatures. Compared with satellite-based observations, UAV data provide much higher spatial and temporal resolution, enabling the detection of localized thermal anomalies that are often overlooked by coarse-resolution satellite imagery. However, several limitations need to be acknowledged. UAV flights are restricted in terms of coverage and endurance, and multiple sorties may be necessary to monitor larger micro-watershed areas over time. Furthermore, intensive calibration is required for emissivity, atmospheric conditions, and ground validation to ensure reliable LST estimates. Although few studies explicitly use the term "micro-watershed," the underlying principles and techniques of UAV-based LST–NDVI modeling can be directly transferred to such environments, offering a valuable tool for evaluating thermal hot spots and vegetation stress in small-scale catchment areas [30].

5.3 Vegetation stress

UAV-based LST–NDVI modeling has become an effective analytical approach for detecting early-stage vegetation stress within micro-watersheds, particularly where the spatial heterogeneity of hydrological conditions plays an important role. High-resolution multispectral and thermal imagery acquired by UAVs allows for the generation of spatially explicit data, such as NDVI and LST, which can be integrated to evaluate the biophysical status of vegetation and identify localized thermal hot spots [18]. Recent case studies in precision agriculture (e.g., vineyard systems) have demonstrated that UAV-derived NDVI values are significantly correlated with in-situ plant water potential measurements, and that including red-edge reflectance bands or LST variables often improves model reliability for site-specific stress prediction. However, the current literature also highlights several methodological limitations of UAV-based LST–NDVI modeling [30]. The radiometric correction, mosaicking, and data fusion required to combine thermal and multispectral imagery are computationally intensive and often require manual intervention to avoid geometric misalignment. Moreover, the modeling outcomes are highly sensitive to environmental conditions, such as variable illumination at different times of day, wind-induced canopy movement, and high reflectance from adjacent soil or water surfaces. These factors introduce noise that can strongly affect the reliability of vegetation stress detection if not adequately accounted for. Finally, the transferability of stress models across different micro-watersheds is limited because vegetation responses are influenced not only by spectral and thermal characteristics but also by local climatic and terrain conditions, meaning that most models must be recalibrated with site-specific field data.

5.4 LU planning

UAV-based modeling of LST and NDVI has emerged as a valuable approach for land-use planning in micro-watersheds. By capturing high-resolution spatial and temporal data, UAVs enable precise mapping of vegetation health, thermal variability, and land cover dynamics. This information supports informed decision-making for sustainable land management, including microclimate regulation, soil erosion control, and vegetation monitoring. Integrating UAV-derived LST and NDVI data allows planners to identify areas vulnerable to environmental stress and optimize LU strategies to maintain ecological balance and productivity [30, 32]. Despite these advantages, UAV-based LST–NDVI modeling has limitations. High-resolution data require substantial computational resources and advanced processing techniques, which can be a barrier to adoption. Additionally, UAVs are constrained by limited flight durations, regulatory requirements, and operational costs, which may restrict repeated measurements or large-scale deployment. These challenges highlight the need for careful planning in UAV missions, including sensor selection, flight scheduling, and integration with other geospatial datasets [65]. Overall, UAV-based LST–NDVI modeling offers a powerful tool for micro-watershed land-use planning by providing detailed environmental insights that support sustainable management practices [13]. When combined with appropriate analytical frameworks and regulatory compliance, UAV data can enhance decision-making for land allocation, conservation strategies, and ecosystem monitoring, ultimately contributing to more resilient and sustainable landscapes [61].

5.5 Water management

UAV-based modeling of LST and NDVI has emerged as a valuable tool for water management in micro-watersheds. By providing high-resolution spatial and temporal data, UAVs enable detailed monitoring of vegetation health and water stress, supporting precise irrigation planning and sustainable water resource management [10]. Integration of LST and NDVI allows for the assessment of soil moisture variability, identification of drought-prone areas, and evaluation of irrigation efficiency through indices such as the CWSI. These capabilities enhance decision-making by enabling timely interventions and optimizing water use in small-scale watersheds [65]. Despite these advantages, UAV-based LST–NDVI modeling faces several limitations. High-resolution data require advanced processing techniques, including radiometric calibration and data fusion, to ensure accuracy. Environmental factors, such as atmospheric conditions, solar angle, and surface emissivity, can introduce measurement errors [40]. Additionally, scaling UAV applications to larger areas is challenging due to flight time constraints, regulatory restrictions, and complex data management. Finally, the cost of UAV platforms and the need for specialized expertise remain significant barriers.

5.6 Integrating appropriate UAV technology into micro-watershed land-use management for a sustainable future

5.6.1 Effects of climate change

Micro-watershed regions are increasingly vulnerable to the effects of climate change, land degradation, and unsustainable land-use (LU) practices. As a response, the integration of UAV technology in micro-watershed land-use management has emerged as a transformative solution, especially when aligned with the principles of appropriate technology affordability, contextual adaptability, and environmental sustainability. This study highlights key components of UAV deployment in micro-watershed contexts, including the use of lightweight UAVs equipped with multispectral and thermal sensors for monitoring vegetation cover, soil erosion, and moisture dynamics [53]. Coupled with open-source geographic information systems (GISs) and real-time communication infrastructure such as 4 G/5 G or LoRaWAN networks, these systems enable high-resolution spatial analysis and rapid decision-making in resource-constrained settings. Furthermore, advancements in artificial intelligence (AI) and swarm UAV coordination enhance the efficiency of land cover classification and environmental forecasting [11]. The integration of UAVs with Internet of Things (IoT) networks and blockchain technologies further strengthens data reliability, transparency, and participatory governance. This chapter argues that appropriate UAV technologies, when embedded within local knowledge systems and digital infrastructures, provide scalable and resilient tools for managing micro-watersheds. The outcomes contribute to ecological balance, sustainable agriculture, and inclusive development, positioning UAVs as key enablers in the transition toward sustainable LU planning. Combined with environmental variables such as slope, roughness, and LU to parameterize hydrological equations, the continuity equation is used to simulate changes in water depth over space and time as a function of rainfall, infiltration, evaporation, and surface flow. The flow velocity in each cell is then estimated using spatially varying Manning's coefficients derived from UAV-based LULC [28, 55].

5.6.2 Appropriate UAV technology for micro-watershed land-use management toward a sustainable future

The integration of UAVs into micro-watershed LU management represents a transformative advancement in sustainable resource governance. To ensure long-term viability, UAV systems must conform to the principles of appropriate technology, affordability, adaptability to local contexts, and environmental compatibility. Essential components include lightweight UAV platforms equipped with multispectral and thermal sensors, enabling the monitoring of vegetation dynamics, erosion patterns, and soil moisture content. The use of open-source GIS facilitates spatial analysis for land-use planning and watershed zoning. Real-time data acquisition, supported by 4 G/5 G or LoRaWAN networks and edge computing, improves operational responsiveness and decision-making efficiency [58]. Emerging trends point toward the integration of AI for automated LC classification and risk forecasting [11]. Swarm UAV deployment allows for synchronized data collection across expansive terrain, while IoT connectivity enhances precision monitoring. Additionally, blockchain-based systems are gaining attention for securing LU data and supporting transparent, decentralized environmental governance. In summary, appropriate UAV technology, when synergized with indigenous knowledge, digital infrastructure, and ecological stewardship, offers a scalable and inclusive approach to managing micro-watersheds. Such integration supports ecological integrity, enhances agricultural productivity, and promotes resilience within rural and resource-sensitive communities [9].

5.7 Emerging trends and technological linkages

Recent advancements have strengthened the role of UAVs in precision watershed management. AI algorithms are increasingly employed to automate land cover classification and predict hydrological risks, while swarm UAV technology enables rapid data acquisition across fragmented or rugged terrain [11]. The convergence of UAVs with IoT sensor networks facilitates multi-source environmental data collection. Additionally, blockchain-based systems are gaining traction for safeguarding geospatial data integrity and enhancing transparency in land governance [65]. These technological integrations position UAVs as central actors in the evolution of smart, sustainable watershed systems. Integrating appropriate UAV technology into micro-watershed land-use management offers a scalable and context-sensitive approach to achieving sustainability goals [61]. When harmonized with local knowledge, geospatial platforms, and participatory governance, UAVs become not only data-gathering tools but also catalysts for ecological resilience, agricultural productivity, and equitable resource planning.

6. Conclusion

In summary, micro-watersheds represent an effective spatial scale for implementing high-resolution land and water management strategies [60]. Their relatively uniform LU characteristics and simpler hydrological structure provide an ideal environment for the application of UAVs-based monitoring tools. UAV platforms equipped with multispectral, thermal, and LiDAR sensors offer substantial improvements in spatial data collection, enabling accurate identification of landscape dynamics, thermal hot spots, and erosion-prone areas at submeter resolution [19]. These fine-scale datasets support the development of spatially explicit hydrological and LU models that can more accurately capture the biophysical processes influencing micro-watershed sustainability [61]. Looking forward, the integration of UAV-derived data with advanced modeling techniques, particularly PIML and graph-based computational frameworks, offers new opportunities for adaptive management and climate-resilient planning in micro-watersheds. However, the operational efficiency and scalability of UAV-based monitoring depend critically on appropriate platform selection, payload transfer efficiency, and energy management strategies. A coordinated approach that combines UAV technology with complementary remote sensing data and participatory monitoring can facilitate more comprehensive, data-driven decision-making. Ultimately, UAV-supported micro-watershed assessment provides a viable pathway for sustainable LU planning, promoting environmental protection and improved resource management in complex and dynamic landscape systems.

Abbreviations/Acronym/Symbol description

AI	artificial intelligence
CNN	convolutional neural network
CWSI	Crop Water Stress Index

DEM	digital elevation models
GIS	geographic information systems
GNN	graph neural network
HUCs	hydrologic unit codes
IoT	Internet of Things
LC	land cover
NDVI	land surface temperature
LU	land use
LU Change	land-use changes
LiDAR	light detection and ranging
NDVI	Normalized Difference Vegetation Index
PIML	physics-informed machine learning
RGB	RED, GREEN, and BLUE light sources
SWAT	soil and water assessment tool
SfM	Structure-from-Motion
SEBAL	surface energy balance algorithm for land
UAVs	unmanned aerial vehicles
UHIs	urban heat mitigation

Author details

Rattana Boonprasert
Faculty of Environment and Resource Studies, Mahidol University, Nakhon Pathom, Thailand

*Address all correspondence to: rattana.boo@mhidol.ac.th

References

[1] Yu X, Duffy C J. Watershed Hydrology: Scientific Advances and Environmental Assessments. Water. 2018;**10: 288**:1–16. DOI: 10.3390/w10030288

[2] Kumar A, Bhardwaj A, Kumar P Padaliya H, et al. Delineation of Micro-Watershed Using Geo-Spatial Techniques. Journal Of Environment, Science And Technology 2017;**3**:14–34

[3] Shekar P.R., Mathew A. Watershed prioritization of drainage basins based on geomorphometric parameters, Neyyar watershed, India. Quaestiones Geographicae. 2023;**42**(3):29–47

[4] Jamwal A K, Thakur P K. In: Chapter 12 The Study of Micro Watersheds for Their Planning, Management, and Implementation Through the Integrated Approaches of Geospatial and Social Sciences: A Case Study from Dharampur Block, Himachal Pradesh, Western Himalayas and Prem Kumar Thakur Modern River Science for Watershed Management. Water Science and Technology Library. 2024;271–239. DOI: 10.1007/978-3-031-547 04-1_12

[5] Raju K S, Kumar D N. Classification of micro watersheds based on morphological characteristics. Journal of Hydro-environment Research. 2011;**5** (2):101–109. DOI: 10.1016/j.jher.2010.09.002

[6] Darama Y, Sey-rek K. Determination of Watershed Boundaries in Turkey by GIS Based Hydrological River Basin Coding. Journal of Water Resource and Protection. 2016;**8**965–981. DOI: 10.4236/ jwarp.2016.811078

[7] Medeiros B M, Cândido B, Jimenez P A J, Avanzi J.C, Silva M L N, et al. UAV-Based Soil Water Erosion Monitoring: Current Status and Trends. Drones. 2025;**9**(4):305. DOI: 10.3390/drones9040305

[8] Chidodo S, Zhang L, Zarei A and Feger K-H, et al. Remote sensing-based assessment of four decades of land use/land cover change in the Sigi River Watershed, East Usambara Mountains, Tanzania. Frontiers in Remote Sensing. 2025;**6**:1594331

[9] Mazzoleni M, Paron P, Reali A, Juizo D, Manane J, Brandimarte L, et al Testing UAV-derived topography for hydraulic modelling in a tropical environment. Natural Hazards. 2020;**103**:139–163

[10] Acharya B S, Bhandari M, Bandini F, Pizarro A, Perks M, Joshi D R, Wang S, Dogwiler T, Ray R L, Kharel G, Sharma S. Unmanned Aerial Vehicles in Hydrology and Water Management: Applications, Challenges, and Perspectives. Water Resources Research. 2021;**57**(11):1–33. DOI: 10.1029/2021WR029925

[11] Chukwu N, Ogwugwa E.C. Applications of Artificial Intelligence in Agriculture: A Review. Engineering, Technology and Applied Science Research. 2019;**9**(4):4377–4383. DOI: 10.48084/etasr.2756

[12] Deng L, Mao Z, Li X, Hu Z, Duan F, Yan Y. UAV-based multispectral remote sensing for precision agriculture: A comparison between

different cameras. ISPRS Journal of Photogrammetry and Remote Sensing. 2018;**14**:124–136. DOI: 10.1016/j.isprsjprs.2018.09.008

[13] Ahmed S, Basu N, Nicholson C.E, Rutter S.R, Marshall, J. R., Perry J. J, Dean J R, et al. Use of machine learning for monitoring the growth stages of an agricultural crop†. Sustainable Food Technology. 2024;**2**:104–125

[14] Freeman P K, Freeland RS. Agricultural UAV in the U.S.: potential, policy, and hype. Remote Sensing Applications: Society and Environment. 2015;**2**:35–42. DOI: 10.1016/j.rsase.2015.10.002

[15] Mirzaee S, C Gomez, Pajouhesh M, Abdollahi K. 1Chapter 16 - Soil erosion and sediment change detection using UAV technology. Remote Sensing of Soil and Land Surface Processes. 2024;271–279. DOI: 10.1016/B978-0-443-15341-9.00025-3

[16] Eltner A. Sofia b G: Chapter 1 - Structure from motion photogrammetric technique. Developments in Earth Surface Processes. 2020;**23**:1–24. DOI: 10.1016/B978-0-444-64177-9.00001-1

[17] Carrivick J L, Smith M W. Fluvial and aquatic applications of Structure from Motion photogrammetry and unmanned aerial vehicle/drone technology. Remote Sensing of Water Environments. 2018;**6**(1):1–17. DOI: 10.1002/wat2.1328

[18] Daniels L, Eeckhout E, Wieme J, Dejaegher Y, Audenaert, K, Maes W.H. Identifying the Optimal Radiometric Calibration Method for UAV-Based Multispectral Imaging. Remote Sensing. 2023;**15**(2909):1–22. DOI: 10.3390/rs15112909

[19] Sankey T T, McVay J, Swetnam T L, McClaran M P., Heilman P Nichols M. UAV hyperspectral and LiDAR data and their fusion for arid and semi-arid land vegetation monitoring. Remote Sensing in Ecology and Conservation. 2018;**4**(1):20–33

[20] Villarreal M L, Bishop T B B, Sankey T T, Smith W K, Burgess M A, Caughlin T. T, Gillan J K, Havrilla, Huang T, LeBeau R L, Norton C L, Sankey J B, Scholl V M, Nonn J, Yao E H, et al. Applications of unoccupied aerial systems (UAS) in landscape ecology: A review of recent research, challenges and emerging opportunities. Landscape Ecology. 2025;**40**(2). DOI: 10.1007/s10980-024-02040-6

[21] Haskins J, Endris C, Thomsen A S, Gerbl F, Fountain M C, Wasson K. UAV to Inform Restoration: A Case Study From a California Tidal Marsh. Frontiers in Environmental Science. 2021;**9**:642906. DOI: 10.3389/fenvs.2021.642906

[22] Awais M, Li W., Hussain S, Cheema M.J.M, Li W, Song R, Liu C. Comparative Evaluation of Land Surface Temperature Images from Unmanned Aerial Vehicle and Satellite Observation for Agricultural Areas Using In Situ Data. Agriculture. 2022;**12**:184. DOI: 10.3390/agriculture12020184

[23] Sankey, Donager J, McVay J, Sankey J B. UAV hyperspectral and LiDAR fusion for forest monitoring in the southwestern USA. Remote Sensing of Environment. 2017;**195**:30–43. DOI: 10.1016/j.rse.2017.04.007

[24] Sivamani S, Kumar G, Gudipalli A. A comprehensive review on payloads of unmanned aerial vehicle. The Egyptian

Journal of Remote Sensing and Space Sciences. 2024;**27**:63–64. DOI: 10.1016/j.ejrs.2024.08.001

[25] Eide A, Koparan C, Zhang Y, Ostlie, M, Howatt K, Sun X, et al. UAV-Assisted Thermal Infrared and Multispectral Imaging of Weed Canopies for Glyphosate Resistance Detection. Remote Sensing. 2021;**13** (4606):1–18. DOI: 10.3390/rs13224606

[26] Boonprasert R, Vijuksungsith P. Agricultural UAV in Recent Advances, Innovations and Applications. In: Advances in Grau A. In: Munguia R, editor. Unmanned Aerial Vehicles Technology and Applications Big Issues Solved with Drone Technology. IntechOpen; London, United Kingdom. 2025. p. 1–31. DOI: 10.5772/intechopen.1010104

[27] Mashala M J, Dube T, Mudereri B T, Ayisi K K, Ramudzuli M R, et al. A Systematic Review on Advancements in Remote Sensing for Assessing and Monitoring Land Use and Land Cover Changes Impacts on Surface Water Resources in Semi-AridTropical Environments. Remote Sensing. 2023;**15** (16):1–29. DOI: 10.3390/rs15163926

[28] Angelini L P, Biudes M S, Machado N G, Geli, H M E., Vourlitis G L, Ruhoff A, Nogueira J D S, et al. Surface Albedo and Temperature Models for Surface Energy Balance Fluxes and Evapotranspiration Using SEBAL and Landsat 8 over Cerrado-Pantanal, Brazil. Sensors. 2021;**21**:7196

[29] McFeeters S K. Using the Normalized Difference Water Index (NDWI) within a Geographic Information System to Detect Swimming Pools for Mosquito Abatement: A Practical Approach. Remote Sensing. 2013;**5**:3544–3561. DOI: 10.3390/rs5073544

[30] Wahab I, Hall O, Jirström M. Remote Sensing of Yields: Application of UAV Imagery-Derived NDVI for Estimating Maize Vigor and Yields in Complex Farming Systems in Sub-Saharan Africa. Drones. 2018;**2**:28. DOI: 10.3390/drones2030028

[31] Elmarakby E, Elkadi H. Prioritising urban heat island mitigation interventions: Mapping a heat risk index. Science of The Total Environment. 2024;**948**:174927. DOI: 10.1016/j.scitotenv.2024.174927

[32] Xu S, Xu X, Blacker C, Gaulton R, Zhu Q, Yang, M, Yang, G, Zhang J, Yang Y, Yang M, et al Estimation of Leaf Nitrogen Content in Rice Using Vegetation Indices and Feature Variable Optimization with Information Fusion of Multiple-Sensor Images from UAV. Remote Sensing. 2023;**15**:854

[33] Vargas V, Carrasco N, Vargas C. Local Participation in Forest Watershed Management: Design and Analysis of Experiences in Water Supply Micro-Basins with Forest Plantations in South Central Chile. Forests. 2019;**10**(7):1–17. DOI: 10.3390/f10070580

[34] Parajuli P B, Ouyang Y. Watershed-Scale Hydrological Modeling Methods and Applications. In: Bradley P M, editor. Current Perspectives in Contaminant Hydrology and Water Resources Sustainability. Rijeka, Croatia: InTech; 2013. p. 58–79. DOI: 10.5772/53596

[35] Colomina I, Molina P. Unmanned aerial systems for photogrammetry and remote sensing. ISPRS Journal of

Photogrammetry and Remote Sensing. 2014;**92**:79–97. DOI: 10.1016/j.isprsjprs.2014.02.013

[36] Evans A D, Gardner K H, Greenwood S., Still B, et al. UAV and Structure-From-Motion Photogrammetry Enhance River Restoration Monitoring: A Dam Removal Study. Drones. 2022;**6**(100): 1–35. DOI: 10.3390/drones6050100

[37] Filho F.H.L, Heldens W.B, Kong Z, Lange E.S. Drones: Innovative Technology for Use in Precision. Pest Management. 2020;1–24. DOI: 10.1093/jee/toz268

[38] Iqbal I M, Balzter H, Firdaus-e-Bareen, Shabbir A. Mapping Lantana camara and Leucaena leucocephala in Protected Areas of Pakistan: A Geo-Spatial Approach. Remote Sensing. 2023;**15**(1020):1–20. DOI: 10.3390/rs15041020

[39] Zhang Y, Wang M, Wang C, Ji L. The Impact Mechanism of Corporate Water-Saving Behaviors on Consumer Response: A Case Study on High Water-Consuming Corporations. Water. 2025;**17**(123):1–25. DOI: 10.3390/w17010123

[40] Klemas V V. Coastal and Environmental Remote Sensing from Unmanned Aerial Vehicles: An Overview. Journal of coastal research. 2015;**315**(5):1260–1267. DOI: 10.2112/JCOASTRES-D-15-00005.1

[41] Rahman M F F, Fan S., Zhang Y, Chen L. A Comparative Study on Application of Unmanned Aerial Vehicle Systems in Agriculture. Agriculture. 2021;**11**:22. DOI: 10.3390/agriculture11010022

[42] Gonzalez-Dugo V, Zarco-Tejada P, Nicola´s E, Nortes P A, Alarco´n J J, Intrigliolo D S, Fereres E, et al Using high resolution UAV thermal imagery to assess the variability in the water status of five fruit tree species within a commercial orchard. Precision Agriculture. 2013;**14**:660–678

[43] Vélez-Nicolás M, García-López S, Barbero L, Ruiz-Ortiz V, Sánchez-Bellón Á. Applications of Unmanned Aerial Systems (UASs) in Hydrology: A Review. Remote Sensing. 2021;**13**(7):1–42. DOI: 10.3390/rs13071359

[44] Zhang H, Wang L, Tian T., Yin, J, et al A Review of Unmanned Aerial Vehicle Low-Altitude Remote Sensing (UAV-LARS) Use in Agricultural Monitoring in China. Remote Sensing. 2021;**13**:1221

[45] Kumar S S G. A Gudipalli: A comprehensive review on payloads of unmanned aerial vehicle. The Egyptian Journal of Remote Sensing and Space Sciences. 2024;**27**(4):637–644. DOI: 10.1016/j.ejrs.2024.08.001

[46] Mohsan S A H, Othman, Khan M A., Amjad N Q H, Zywiołek J A. Comprehensive Review of Micro UAV Charging Techniques. Micromachines. 2022;**13**(6):1–30. DOI: 10.3390/mi

[47] Manfreda S, McCabe M F, Miller P E, Lucas R, Madrigal V P, Mallinis G, Dor E B, Helman D, Estes L, Ciraolo G, Müllerová J, Tauro F, Lima M I, Lima J L M P, Maltese A, Frances F, Caylor K, Kohv M, Perks M, Ruiz-Pérez G, Su Z, Vico G, Toth B, et al On the Use of Unmanned Aerial Systems for Environmental Monitoring. Remote Sensing. 2018;**641**:1–25

[48] Impollonia G, Croci. M, Ferrarini A, Brook J, Martani E, Blandinières H,

Marcone A, Awty-Carroll D, Ashman C, Kam J, et al. UAV Remote Sensing for High-Throughput Phenotyping and for Yield Prediction of Miscanthus by Machine Learning Techniques. Remote Sensing. 2022;**14**(2927):1–24. DOI: 10.3390/rs14122927

[49] Abdulkadirov R, Lyakhov P, Butusov D,Nagornov N, Kalita D, et al. Physics-Aware Machine Learning Approach for High-Precision Quadcopter Dynamics Modeling. Drones. 2025;**9**(187):1–26. DOI: 10.339 0/drones9030187

[50] Tsunetaka H, Hotta N, HayakawaY S, Imaizumi F, et al. Spatial accuracy assessment of unmanned aerial vehicle-based structures from motion multi-view stereo photogrammetry for geomorphic observations in initiation zones of debris flows, Ohya landslide, Japan. Progress in Earth and Planetary Science. 2020;7(24):2–14. DOI: 10.1186/ s40645-020-00336-0

[51] Carrivick J L, Smith M W. Fluvial and aquatic applications of Structure from Motion photogrammetry and UAV/drone technology . Wiley Interdisciplinary Reviews: Water. 2018;**6**:1. DOI: 10.1002/wat2.1328

[52] Wang J, Ma C, Chen P, Yao W, Yan Y, Zeng T, Chen S and Lan Y, et al. Evaluation of aerial spraying application of multi-rotor unmanned aerial vehicle for Areca catechu protection. Frontiers in Plant Science. 2023;**14**:1093912. DOI: 10.3389/fpls.2023.1093912

[53] Anderson K, Gaston K.J. Lightweight unmanned aerial vehicles will revolutionize spatial ecology. Frontiers in Ecology and the Environment. 2013;**11**(3):138–146. DOI: 10.1890/120150

[54] Bhandari M, Xue Q, Liu S, Stewart B. A, Rudd J. C, Pokhrel P, Blaser B, Jessup K, Baker J. Thermal imaging to evaluate wheat genotypes under dryland conditions. Agrosystems, Geosciences & Environment. 2021;**4**: e20152. DOI: 10.1002/agg2.20152

[55] Yuana G, Wanga Y, Zhaoa F, Wanga T, Zhanga L, Haoa M. Accuracy assessment and scale effect investigation of UAV thermography for underground coal fire surface temperature monitoring. International. International Journal of Applied Earth Observation and Geoinformation. 2021;**102**:102426. DOI: 10.1016/j. jag.2021.102426

[56] Yan W, Jiang J, He Lanwu, Zhao W, Nair R, Wang X. Correcting land surface temperature from thermal imager by considering heterogeneous emissivity. International Journal of Applied Earth Observation and Geoinformation. 2024;**129**:103824. DOI: 10.1016/j.jag.2024.103824

[57] Ezzy H, Charter B. M, Brook A. How the Small Object Detection via Machine Learning and UAS-Based Remote-Sensing Imagery Can Support the Achievement of SDG2: A Case Study of Vole Burrows. Remote Sensing. 2021;**13**(3191):1–18. DOI: 10.3390/ rs13163191

[58] Alexandris S, Psomiadis E, Proutsos N, Philippopoulos P, Charalampopoulos I, Kakaletris G, Papoutsi E-M., Vassilakis S, Paraskevopoulos A, et al. Integrating Drone Technology into an Innovative Agrometeorological Methodology for the Precise and Real-Time Estimation of Crop Water Requirements. Hydrology. 2021;**8**:131

[59] Jeziorska J. UAS for Wetland Mapping and Hydrological Modeling. Remote Sensing. 2019;**11**(17):1997. DOI: 10.3390/rs11171997

[60] Zarco-Tejada P J, Hornero A, Hernández-Clementec R, Beck P S A, et al Understanding the temporal dimension of the red-edge spectral region for forest decline detection using high-resolution hyperspectral and Sentinel-2a imagery. ISPRS Journal of Photogrammetry and Remote Sensing. 2018;**137**:134–148

[61] Fang X, Li J, Ma Q. Integrating green infrastructure, ecosystem services and nature-based solutions for urban sustainability: A comprehensive literature review. Sustainable Cities and Society. 2023;**98**(104843):1–13. DOI: 10.1016/j.scs.2023.104843

[62] Lost Filho F. H, Heldens W.B, Kong Z, de Lange E, et al Drones: Innovative Technology for Use in Precision Pest Management. Journal of Economic Entomology. 2020;**113**:1–25

[63] Saravanakumar Y N, Sultan M T H, Shahar FS, Giernacki W, Łukaszewicz A, Nowakowski M., Holovatyy A, Stepien´ S, et al. Power Sources for Unmanned Aerial Vehicles: A State-of-the Art. Applied Sciences. 2023;**13**(21):1–27. DOI: 10.3 390/app132111932

[64] Li H, Li C, Lin Y, Lei Y. Surface temperature correction in TVDI to evaluate soil moisture over a large area. Journal of Food, Agriculture & Environment. 2010;**8**(3&4): 1141–1145

[65] Kazemi F, Parmehr GE. Evaluation of RGB vegetation indices derived from UAV images for rice Crop Growth Monitoring. In: ISPRS Annals of the Photogrammetry, Remote Sensing and Spatial Information Sciences; 19–22 February 2023. Tehran, Iran (virtual): GeoSpatial Conference 2022 (X-4/W1-2022). p. 385–390. DOI: 10.5194/isprs-annals-X-4-W1-2022-385-2023